Molecular Biology of the Biological Control of Pests and Diseases of Plants

Edited by

Muthukumaran Gunasekaran
Department of Biology
Fisk University
Nashville, Tennessee

and

Darrell J. Weber
Department of Botany and Range Science
Brigham Young University
Provo, Utah

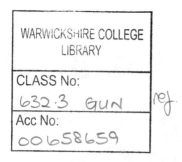
Library of Congress Cataloging-in Publication Date

Molecular biology of the biological control of pests and diseases of
    plants / edited by Muthukumaran Gunasekaran and Darrell J. Weber.
        p.   cm.
    Includes bibliographical references and index.
    ISBN 0-8493-2442-4 (alk. paper)
    1. Pests—Biological control.   2. Plant diseases.
I. Gunasekaran, Muthukumaran, 1942–   .   II. Weber, Darrell J.,
1933–   .
SB975.M64   1995
632'.96—dc20   95-31087
                                                                 CIP

Pershore College

PREFACE

Biological control methods have been utilized since the first day man began to fight plant pests and diseases. Early biocontrol strategies involved crop rotation to avoid plant pests, as well as cross breeding plants to create more resistant plants. With the discovery of chemical pesticides and herbicides, biological control strategies were placed on the shelf, made obsolete by the new technology. However, after decades of use, we have begun to realize the inadequacies of our technology. Chemical pesticides have proven to be very economical and convenient, but they have had some serious drawbacks. They have proven to be less effective after repeated use since many insects have developed a resistance to the agents. Furthermore, chemical pesticides have polluted our soil, streams, rivers, and other water sources, introducing chemical agents into our environment that are harmful to plants, animals, and humans.

Confronted with these shortcomings, biological control methods have been taken off of the shelf again. Recent advances in molecular biological techniques offer the possibility of making more effective and practical biological control methods. Biological control strategies typically consist of strategies to make the crop plant resistant to pathogens or to antagonize the plant pathogens. New biotechnology has improved both of these strategies.

Crop plants are constantly under attack from a wide range of plant pathogens. There are different biocontrol strategies to combat these different pests. Each chapter in this book deals with a different pathogen and the application of new molecular biological techniques to the biocontrol of the pathogen. In the introductory chapter, we discuss the types of control methods that are used. The techniques commonly used in molecular biology to identify the etiological agents, diagnose various diseases, and develop the control methods of pests and diseases are reviewed. In Chapter 2, Grumet and Lanina-Zlatkina discuss the new approaches to the control of viral diseases in plants. They include resistance related to the coat protein gene, replicase genes, and movement proteins genes. The new techniques involve creating resistant plants as well as approaches to utilize viral parasites. In Chapter 3, Sands and Pilgeram discuss novel genetic approaches in the control of plant parasitic bacteria. In Chapter 4, Fulbright and Smart discuss several applications of molecular biology in the biological control of fungal diseases. In Chapter 5, Stirling, Eden, and Aitken address the control of nematodal diseases in plants. They suggest that molecular biology can help in the taxonomic problems of nematodes as well as determine the variability within species. However, more information is needed on the mechanism of action of many of the biological control agents before manipulations of useful genes can occur. In Chapter 6, Wood reviews the use of baculoviruses in the control of plant parasitic insects. He comments that the new molecular approaches will result in recombinant pesticides partly because it costs less to develop and register a baculovirus pesticide. Six baculovirus pesticides have already been registered. In Chapter 7, Dean and Rajamohan address the molecular biology of bacteria in the control insects. They highlight the success of the molecular approach using the toxins from *Bacillus thuringiensis* and *Bacillus sphaericus* as insecticidal agents. In

Chapter 8, Clarkson discusses the use of insect pathogenic fungi to control insects. He discusses the role of endoproteases and aminopeptidase in relation to appressoria formation. Toxins from fungi, especially the cyclic depsipeptides, appear to have potential as biological insecticides. In Chapter 9, Heckmann discusses the use of protozoa as biological control agents. Kennedy in Chapter 10 reviews the impact of recent advances in molecular biology on the development of bioherbicides to control weeds. She highlights the particular effectiveness of bacteria and fungi in this task. In Chapter 11, Ragunathan and Divakar narrate the development of integrated pest management in developing countries combating the major pests and diseases of important crop plants. In Chapter 12, we review the successful examples of biological control and suggest future trends and research needs.

At this point, very few biocontrol agents are commercially available to control plant pests and diseases. What this work serves to show is that much of the foundational work has already been done for their production, and with the proper application of recently developed techniques, biological control methods will not represent the past, but the future of controlling plant pest and diseases safely and effectively without harming the environment and human health.

Muthukumaran Gunasekaran
Darrell Jack Weber

THE EDITORS

Muthukumaran Gunasekaran, Ph.D., is a Professor in the Department of Biology at Fisk University in Nashville, Tennessee.

Dr. Gunasekaran graduated with a B.S. Degree in Agriculture in 1964 from Annamalai University, Tamil Nadu, India. In 1966, he received his M.S. Agriculture degree in Plant Pathology from the University of Madras, India. He obtained his Ph.D. in Plant Pathology in 1970 from Texas A & M University, College Station, Texas. He was a post doctoral fellow in botany at the Brigham Young University, Provo, Utah from 1970–1972. Subsequently, he was a faculty member in the department of Infectious Diseases at St. Jude Children's Research Hospital from 1973–1981, where he worked in clinical mycology. From 1981 to present, he has been professor of biology at Fisk University. In 1990, he was a Fulbright Teaching Fellow in the Department of Botany, Bharathidasan University, Trichy, India.

Dr. Gunasekaran is a member of the American Mycological Society, the Indian Phytopathological Society, the Society of Industrial Microbiology, and the American Society for Microbiology.

Dr. Gunasekaran has presented over 80 research papers at National and International meetings and has published more than 80 research papers. He has received research grants from the National Institutes of Health, the American Cancer Society, the Howard Hughes Medical Institute, the National Science Foundation, the Department of Education, the Department of Energy, the National Aeronautic Space Agency, the U.S. Forest Service, and the U.S. Department of Agriculture. His major research interest is biochemistry and physiology of host-parasite relationships and regulation of fungal dimorphism.

Darrell Jack Weber, Ph.D., is a Professor of Botany in the Department of Botany and Range Science at Brigham Young University in Provo, Utah.

Dr. Weber graduated in 1958 from the University of Idaho in Moscow, Idaho with a B.S. degree in Agricultural Biochemistry. In 1961, he received his M.S. degree in Agricultural Biochemistry from the University of Idaho. He received his Ph.D. in Plant Pathology from the University of California at Davis, California in 1963. He was a post-doctorate in biochemistry at the University of Wisconsin in Madison, Wisconsin from 1963–1965. He spent a sabbatical leave in the Department of Biochemistry at Michigan State University, East Lansing, Michigan during 1975–1976. In 1995, he was a Fulbright Research Fellow in the Department of Botany of the University of Natal in Pietermaritzburg, South Africa.

Dr. Weber is a member of Alpha Zeta, Sigma Xi, the American Mycological Society, the American Phytopathological Society, the Association for the Advancement of Industrial Crops, the North American Mushroom Society, the Utah Native Plant Society, and the Utah Mushroom Society.

Among other awards, he is a fellow of the Utah Academy of Science, and he received the Carl G. Maeser Research Award and a Fulbright Research Fellowship Award.

Dr. Weber has presented over 70 research lectures at National and International meetings. He has published 175 research papers. He is an editor or co-editor of three books. He has been the recipient of research grants from the National Institutes of Health, the National Science Foundation, the U.S. Forest Service and the U.S. Department of Agriculture. His current research interest is biochemistry of plant disease reactions, stress reactions in halophytes, and lipid components of fungi.

CONTRIBUTORS

Elizabeth A. B. Aitken, B.Sc., Ph.D.
Department of Botany
The University of Queensland
Queensland, Australia

John Michael Clarkson, B.Sc., Ph.D.
Microbial Pathogenicity Group
School of Biology and Biochemistry
University of Bath
Bath, England

Donald H. Dean, Ph.D.
Department of Biochemistry
The Ohio State University
Columbus, Ohio

B. Julius Divakar, Ph.D.
Senior Entomologist
Ministry of Agriculture
National Plant Protection Training
  Institute
Hyderabad, India

Lois M. Eden, B.Sc.
Department of Botany
The University of Queensland
Queensland, Australia

Dennis W. Fulbright, Ph.D.
Department of Botany and Plant
  Pathology
Michigan State University
East Lansing, Michigan

Rebecca Grumet, Ph.D.
Department of Horticulture
Michigan State University
East Lansing, Michigan

**Muthukumaran Gunasekaran,
  Ph.D.**
Department of Biology
Fisk University
Nashville, Tennessee

Suresh Gunasekaran, B.S.
Department of Biology
Fisk University
Nashville, Tennessee

Richard A. Heckmann, Ph.D.
Department of Zoology
Brigham Young University
Provo, Utah

Ann C. Kennedy, Ph.D.
USDA-Agriculture Research
  Service
Pullman, Washington

Tatiana Lanina-Zlatkina, Ph.D.
Department of Horticulture
Michigan State University
East Lansing, Michigan

Alice L. Pilgeram, Ph.D.
Department of Plant Pathology
Montana State University
Bozeman, Montana

**V. Ragunathan, Ph.D., F.P.P.A.I.,
F.P.S.I.**
Plant Protection Adviser to the
  Government of India
Ministry of Agriculture
Directorate of Plant Protection,
  Quarantine, and Storage
Faridabad, India

Francis Rajamohan, Ph.D.
Department of Biochemistry
The Ohio State University
Columbus, Ohio

David C. Sands, Ph.D.
Department of Plant Pathology
Montana State University
Bozeman, Montana

Christine D. Smart, Ph.D.
Department of Plant Pathology
University of California
Davis, California

Graham R. Stirling, M.Ag.Sc., Ph.D.
Department of Primary Industries
Plant Protection Unit
Queensland, Australia

Darrell Jack Weber, Ph.D.
Department of Botany and Range
 Science
Brigham Young University
Provo, Utah

H. Alan Wood, Ph.D.
Plant Protection Program
Boyce Thompson Institute
Ithaca, New York

TABLE OF CONTENTS

1 Approaches to the Control of Pests and Diseases of Plants

Darrell Jack Weber and
Muthukumaran Gunasekaran

TABLE OF CONTENTS

0-8493-2442-4/96/$0.00+$.50

I.   IMPACT OF DISEASES AND PESTS

Food surpluses and food shortages occur at the same time in the world today. However, food shortages are a more serious matter. Several famine relief efforts have been undertaken to help in the impoverished areas, particularly in Africa. It is estimated that 800 million people are undernourished, and as many as 2 billion may be suffering from hunger or malnutrition. With the world population near 6 billion and still growing, the question of food production becomes an important concern. Directly related to food production are losses caused by pests and diseases. It is estimated that for the entire world, the percent of crop loss due to diseases, insects, and weeds is about 34%.[1] The estimated percent loss caused by each individual group is 11.9% by diseases, 12.3% by insects, and 9.8% by weeds.[2] An additional 15% food loss could be added due to postharvest pests and diseases.[1] This suggests that the total impact of pests and diseases results in the loss of almost half of the food produced in the world. Effective control of pests and diseases would result in a tremendous increase in food. The amount of loss is less in the developed nations as compared to the developing nations. Ironically, developing nations have a much greater number of their population (57%) engaged in agriculture, but their losses are still higher. Much of the success in controlling pests and diseases in developing nations has been due to the application of large quantities of chemical pesticides. Many of the pesticides are no longer effective due to the development of resistance to the chemical by the pest. In other cases the pesticide has been barred from use because it did not pass reregistration. There is increased concern about using pesticides in the environment because of potential contamination of ground water and foodstuffs.[3] For developing nations, pesticides are a major expense in crop production. On the other hand, pesticides are the most effective means of controlling pests and diseases.

Alternate nonpesticide methods of controlling pests and diseases would be very desirable. Biological control, the suppression or destruction of pests and diseases with living organisms, has the appeal of a reliable and safe alternative to high pesticide use. Biological control is not a new concept. It has been around a long time. There is a record of the ancient Chinese using the ant, *Oecophylla smaragdina,* to control caterpillars and beetles on citrus trees in 324 B.C. In 1752, Linnaeus suggested that insects could be used to control other insects.[4] In 1884, Krassilstschick[5] grew the fungus *Metarrhizium* and used the spores to control the sugar beet curculio larvae (*Cleonus punctiventris*).[4] The first great success in classical biological control of insects occurred in 1889 in the control of cottony cushion scale in California by using the parasitic fly, *Crytochetum iceryae.*[4]

Biological control of weeds started in 1863 with the use of the cochineal insect, *Dactylopius ceylonicus,* to control prickly pear cactus in India. The control of Klamath weed in California in 1946 by the introduction of *Chrysolina quadrigemina*

(Suffrian) was a successful application of biological control. It is estimated that the abundance of Klamath weed was reduced by 99%.[6]

The strategy for obtaining biological control of pests normally involves three approaches using natural enemies: (1) Importation of exotic species and their establishment in a new habitat; (2) augmentation of established species through direct manipulation of their populations by mass production and periodic colonizations; (3) the conservation of natural enemies by manipulations of the environment.[4]

Biological control of plant disease involves the use of antagonistic microorganisms that interact with the parasite to reduce its effectiveness. In the widest sense, biological control of plant diseases can include crop rotation, direct addition of microbes antagonistic to pathogens, use of chemicals to change the microflora, and plant breeding to develop resistant plants.[7] A range of success in biological control of plant diseases has been obtained. Perhaps most effective has been the breeding of resistance plants.

II.   HOST AND PARASITE RELATIONSHIPS

A parasite can be defined as an organism that lives on or in some other organism and obtains its food from the host organism. The interaction between the host and the pathogen is not a passive process but rather an active interaction. The weapons of the pathogens include the ability to physically penetrate the host cell, the secretion of degraded enzymes, production of microbial toxins that weaken or damage the host, synthesis of polysaccharides that can plug the water flow in the xylem of the host, and production of auxins that stimulate the host to respond to the pathogen.[8]

On the other hand, plant hosts protect themselves against pathogen attack by: having preexisting physical structures that prevent the entry of the pathogen, forming protective cork layers when infected, forming abscission layers around the infection area to isolate the pathogen, producing tyloses and gums that localize and restrict the movement of the pathogen, synthesizing of phytoalexins that are inhibitory to pathogens, producing phenols which are normally toxic to the pathogen, detoxifying fungal toxins by enzymatic action, and sacrificing some of their own cells which also kill the pathogen through the hypersensitive reaction.[8]

The interaction between the host and the pathogen is a dynamic process. In biological control, another organism which is antagonistic to the pathogen and favorable to the host is involved in the host–parasite interaction.

Insects can directly penetrate the host or consume part of the host. In some cases, the insect injects chemicals similar to plant hormones into the host which stimulate the host plant to produce galls and witches brooms.

While insects are very mobile, they often induce reactions similar to those of pathogens in the host. Host plants produce phytoalexins, phenolic complexes, and chemicals that are anti-feedants for insects.

III. HOST RESISTANCE

Over the centuries, plants have evolved resistance to many pathogens and pests. When new pathogens and pests are introduced or strains of existing pathogen and pests develop, if no natural resistance is present then major losses will occur in the crops. Plant breeding is a means of adding resistance to the crop plant in response to the presence of serious pathogens and pests.

Plant breeding for resistance in the host has many advantages. The resistant plant can be grown by culture methods similar to those already in practice. There is no need to apply pesticides to the environment. Because no pesticides are used, it is often less expensive to grow the crop.

Among the disadvantages of breeding for resistance is that it takes a number of years to develop the resistant variety. There must be a genetic resistant characteristic to the pathogen of concern in a plant that can be bred with the crop to develop a resistant variety. In many cases, it is not possible to find such resistance in another plant that can be cross-bred. In other cases, the genetic cross can be made but the hybrid is not fertile and the breeding program cannot continue. It requires many backcrosses to maintain the desirable agronomic characteristics in the plant while the resistant characteristics are being incorporated into the genetic system of the host plant.

Once a resistant variety has been developed, the parasite will often evolve and develop a strain that can break the resistance of the host. In many cases, this breakdown in resistance occurs over a long period of time. In other cases, the resistance may be lost quite quickly.

IV. APPROACH TO CONTROL OF PLANT DISEASES AND PESTS

The control of plant diseases and pests has been a constant battle over a long period of time. A number of control approaches have been used.[9]

A.   Physical Control Methods

1. **Heat treatment.** In some cases, control has been obtained by treating the host products, such as fruits, tubers, and seeds, with heat to eradicate any pathogens or pests that may be present. The disadvantage is that the conditions must be carefully controlled because too much heat can damage the plant product.
2. **Sterilization.** Soil sterilization has been another approach to control soil pathogens and pests. Generally chemicals are used which result in killing the pathogens and pests. The chemicals also kill nonpathogens, resulting in changes in the ecology of the soil microflora. Favorable results have been obtained with soil sterilization in growing strawberries. The disadvantage of the soil treatment is that any new pathogen or pest that may come in after the soil sterilization would have little competition in becoming well established because there is an absence of other competing soil microflora.

3. **Radiation.** Another approach to controlling postharvest pathogens and pests is the radiation of fruits, tubers, and seeds. While radiation can destroy the pathogens and pests, it also can damage the host tissue. Radiation has not been a very successful approach.

B.   CHEMICAL CONTROL METHODS

The most common approach to controlling pathogens and pests has been to apply chemicals (pesticides). Contact and systemic sprays are the two most common types used.

1. Contact chemical sprays kill or inhibit the pathogen or pest upon contact. A powder form is often used to inhibit or kill the fungus, bacterium, or insect. It is also common to spray with a liquid solution at regular intervals to control the pathogens or pests. This approach adds chemicals to the environment, and often the pathogen or pests develop resistance to the chemical.
2. Systemic chemical sprays move throughout the plant after they are applied. Even though the pathogen or pest is not present at the time of the spray, it is inhibited or killed because the chemical moves throughout the plant and provides protection against pathogens and pests. This is a more effective control method. However, it is usually more expensive and it does add chemicals to the environment

C.   BIOLOGICAL CONTROL

Biological control, from a theoretical point of view, has a lot of appeal because it does not add chemicals to the environment. The major challenge has been to make it an effective control method and, at the same time, maintain the balance so the biological control agent does not have to be added each time.

Classical methods of biological control include modifying the soil macroflora through crop rotations. The soil microflora can also be changed by adding products like straw to the soil to favor certain microorganisms.

The introduction of predators that will attack the pathogens has had some dramatic success. In the case of insect pests, an example would be the spraying of insects with the bacterium, *Bacillus thuringiensis* Berliner, which results in the death of the insect.[10] In another type of biological control, the host is sprayed with a nonpathogenic organism, such as a noninfective virus, which stimulates the defense reaction of the host plant. This results in a type of cross protection against an attack by a later virus. Certain biological organisms, such as fungi and nematodes, have been applied to weeds, resulting in the reduction of the weed population.[11]

While classical biological control offers potential nonchemical control of pathogens and pests, it has been limited by several problems. It is not always possible to find a biological control organism for the disease or pest to be controlled. The biological control organism may not be specific enough for the disease to be controlled. In some cases, the biological control organism attacks too wide a range of organisms. The environment may not be favorable for the maintenance of a population of the biological control organisms. The genetic characteristic for biological control (resistance) may not be found in a genetically compatible plant.

With the development of molecular biology and biotechnology, methods have been developed that could potentially overcome many of these limitations. For example, some of the resistant genetic components can be transferred between plants that cannot be crossbred.

V.   MOLECULAR BIOLOGY APPROACH: HISTORICAL DEVELOPMENT

The development of molecular biology techniques to obtain and handle DNA has provided the tools for transferring and evaluating genetic characteristics from many different organisms.

Methods have been developed to isolate and purify DNA from most organisms.[12] The development of restriction endonucleases provides the tools to cut the isolated DNA from different organisms at specific sites.[13] These cut DNA segments can be joined together in a plasmid. Cohn's production of recombinant DNA in 1973 with plasmid pSC101 provided the means for transferring sections of DNA from another organism to *E. coli*. The transferred DNA can then be expressed in the *E. coli* cell.[13] Agarose gel electrophoresis provides the means for separating the DNA segments of different molecular weight by means of their electrophoretic mobility. A fluorescent dye, ethidium bromide, binds to the DNA bands which can then be visualized under ultraviolet light. The different bands represent DNA of different molecular weights.[12]

In many cases, the DNA segments obtained by restriction endonucleases is sufficient for agarose gel electrophoresis and for transfers with plasmids. In other cases, the DNA fragments are present in too low a concentration. To overcome this problem, the polymerase chain reaction (PCR), a method of duplicating DNA segments, was developed.[14]

A.   The PCR Technique

Polymerase chain reaction (PCR) is a technique by which any piece of DNA can be amplified in a test tube.[15] This means that a trace amount of DNA can be duplicated repeatedly to provide a sufficient quantity for many different DNA analyses to be performed. The PCR mixture contains a DNA polymerase, an enzyme that is stable at 92–95°C. The DNA polymerase duplicates the short segments of DNA at 37°C; then the temperature is raised to around 94°C to denature the newly formed DNA, which causes it to be released from the template DNA. Next, the mixture is cooled down and the new and old DNA both act as templates for the DNA polymerase to make more DNA. This is a geometric process and, after several hours of cycling high and low temperatures, a considerable amount of DNA is produced.[15]

With the PCR procedure, DNA segments can be produced in sufficient amounts for detailed DNA analysis. It would be desirable to associate the different DNA segments with specific sites in the genome. One approach would be to compare known DNA segment with the unknown DNA segments. The restriction fragment length polymorphism technique (RFLP) is the method used to obtain reference information on the DNA segments.

B. THE RFLP TECHNIQUE

Restriction fragment length polymorphisms (RFLPs) is a powerful method for the detection of specific fragments of DNA. The procedure involves first extracting the DNA from the tissue of an organism. Different DNA restriction enzymes are added to different DNA samples in separate tubes. Each restriction enzyme cuts the DNA in specific locations. The mixture of restriction fragments for each enzyme is submitted to agarose gel electrophoresis to separate them out according to their molecular weight. Each restriction fragment mixture has a specific number of bands according to its molecular weight. The number of bands are normally different for the different endonucleases used. After separation with agarose gel electrophoresis, a radioactive DNA probe (previously prepared from a specific known fragment of DNA) is added, and it binds to the specific fragment with identical base pairing. The base sequence of the individual DNA probe can be determined by gene sequencing and other DNA analyses. It is not possible to see with the normal eye which band the radioactive DNA probe has bound. After the radioactive probe is added to the gel, a sheet of photographic film is laid over the top of the gel and the film is exposed to the radioactivity from the probe bound to the specific band. The film is developed and used to the locate the band with the radioactive probe in relation to the separated restriction fragment bands. This procedure permits the identification of a specific DNA fragment. These identified sites serve as genetic markers and are used to associate certain genes with the location of the marker.[16] A number of nucleic acid probes in combination with PCR amplification have been developed for the detection and identification of pathogens and pests.[16] This technique has become especially useful in establishing pathogen-free and pest-free breeding stock for doing DNA transfer.

C. THE RAPD TECHNIQUE

Another method of obtaining markers for evaluation of the genetic information of organisms is random amplified polymorphic DNA (RAPD).[17] It differs from RFLP in that primers are used instead of the restriction endonucleases. RAPD is a rapid and efficient means for identifying the degree of genetic varibility among accessions in biological resource collections.[15] The procedure involves the repeatable amplification of DNA fragments, using primers of arbitrary sequence. RAPD amplification is based on the ability of a single DNA primer composed of 9–10 nucleotides to determine the amplification of several DNA fragments from a genomic DNA source.

The arbitrary primers recognize specific sites in the genome that have a degree of homology to the primer. In order for a DNA fragment to be amplified, the primer must recognize and bind to a DNA sequence on one strand and a similar sequence within 2000 base pairs on the complementary strand. The PCR technique is used to amplify the DNA seqments produced by the primers. The primer fragments of DNA are separated by agarose gel electrophoresis. Any genetic difference that allows a primer to bind in one genotype but prevents it from binding at that site in another genotype will result in a polymorphism (presence or absence of a DNA fragment). Since presence vs. absence is the primary type of polymorphism, RAPD markers are typically inherited as dominant Mendelian alleles. RAPD markers have bee used in marker-assisted selection,[17] quantification of genetic diversity,[18] paternity used for identification, genome mapping,[18] and genomic fingerprinting.[19] RAPD has been used to characterize genera, species, and races of pathogens.[20] Polymorphisms can be recognized as the presence of a specific amplified DNA fragment from one accession in comparison to the same fragment from another accession. The RAPD procedure is more rapid than RFLP and the procedure has been automated. RAPD has the potential of being used with virtually any species with little method modification.

With methods available for using markers to identify genes, for obtaining DNA segments for the duplication of the DNA sequences, and for adding DNA segments to vectors such as plasmids, the next step is to add the resistance factors that are associated with biological control.

D. TRANSFORMATION TECHNIQUES

Transformation involves the taking up of external DNA and incorporating it into the genome of the host. Transformations in bacteria can be accomplished by putting a segment of DNA in a plasmid. The plasmid is then transferred into a host cell. Eventually, the external DNA segment in the plasmid is incorporated into the host genome.[21] With bacteria, this process has been worked out quite well. However, transformation of DNA in plant cells is more difficult. Part of the difficulty is due to the long generation times, large plant genomes, and the plant cell wall. The Ti plasmid from the plant tumor-inducing bacterium, *Agrobacterium tumefaciens* Erw. Smith and Townsend, is used most commonly as a vector in gene transfer work with plants.[20] In some cases, plant protoplasts or callus tissue are produced to help the Ti plasmid enter the plant cell. Other methods of putting DNA into a plant cell include microinjection, electroevaporation, and particle bombardment.[12,22] However, these methods will not work for all plants. It is more difficult to obtain DNA transfers in monocot plants. With insects, the baculoviruses can be used as vectors to transfer foreign genes into host cells.[23]

With the ability to add genes and modifying genetic characteristics in plants and pathogens, it became apparent that biological control methods could be developed in many different ways that were not previously possible.[24]

VI. CHANGING THE RESISTANCE OF THE HOST WITH TECHNIQUES USED IN MOLECULAR BIOLOGY

A. TRANSFERRING RESISTANT PATHOGEN GENES INTO HOST PLANTS

While classical plant breeding provides a means of incorporating resistance against pests and pathogens, the resistant gene has to be found in a species that can be bred with the crop involved. With the molecular biology approach, potentially resistant genes in a similar plant that was incompatible for breeding with the crop plant can be removed, transferred by a vector to the crop plant, and incorporated with the plant's own genome. For example, Loesch-Fries[25] incorporated the genes for the protein coat of alfalfa mosaic virus into a host plant. The production of the virus protein coat stimulated the host defense system and cross protection against later virus infection was obtained. Beachy et al.[26] transferred the gene for the protein coat of tobacco mosaic virus (TMV) and obtained either delayed symptoms or an absence of symptoms in tobacco, tomato, and potato plants. They also found that the protein coat for TMV also conferred resistance to tomato mosaic virus.

The bacterium, Bacillus *thuringiensis* Berliner, which produces insecticidal crystal proteins (ICP), has been used in biological control as a direct spray on a number of caterpillars including the tobacco hornworm and the gypsy moth. The ICP are converted by enzymatic action to a toxin which kills the insect by generating pores in its membranes.[27] The ICP are considered effective environmentally safe insecticides. The concept is to incorporate the ICP gene into the host plants and have the toxin gene expressed in the host plant to give protection against several types of caterpillar. The toxin gene of *B. thuringiensis* has been introduced into plants and has provided insect resistance.[28]

B. INSERTION OF RESISTANT GENES FROM ANOTHER TYPE OF PLANT SPECIES

If resistant genes can be found towards a particular pathogen or pest, then with molecular biology methods, it should be feasible to transfer the genes, even if they are not compatible for breeding. The gene could be from the same species of plant or from another species. These species may be completely unrelated and the gene transfer may provide protection. The DNA segment containing the resistant characteristic is cut out and put into bacteria with a plasmid. Then the plasmid inserts the DNA segment into the plant's genome. If sucessful, the result would be a resistant variety. Viral resistant plants have been obtained by using antisense RNA.[29] Fungal resistance has been obtained by introduction of a plant or bacterial chitinase gene into a plant.[30]

C. The Mechanism of Gene Transfer Resistance

The genetic characteristic may involve structural change such as thicker cell walls that are difficult for a fungus to penetrate. It could involve the production of a metabolite that acts as a phytoalexin to inhibit the fungus, thereby providing the resistance that was missing before.

VII. CHANGING THE PARASITE WITH TECHNIQUES USED IN MOLECULAR BIOLOGY

In some cases, it is possible to modify the pathogen itself or modify a microbial organism that interacts with the pathogen so as to have a impact on the pathogen. The net effect is a type of biological control that does not involve modifying the host plant.

One such example involves putting the endotoxin gene from *B. thuringiensis* into another bacterium *Pseudomonas fluorescens* Flugge which is normally associated with roots of plants. The transformed *Ps. fluorescens* then secrets the endotoxin that can kill the root cutworm. The net effect is that *Ps. fluorescens* protects the host plant against cut worm damage.[31]

Foreign genes can be incorporated into baculoviruses which then can be used as vectors to transfer the foreign gene into the host cell.[21] Examples of foreign genes are those that cause water loss in insects,[32] those that cause changes in the nature of the juvenile hormone esterase,[33] and those that inactivate insect steroid hormones.[34] Genes for several toxins, such as BT toxin,[35] have been inserted into insect host cells using baculoviruses as vectors. Bishop and Possee[36] altered baculoviruses and tested them in the field as a type of biological control insecticide. They added markers to the baculoviruses in order to follow their field persistence. Baculoviruses attack only arthropods but they are slow in their rate of kill. The baculoviruses were effective in killing a range of caterpillars and did not persist in the environment.[37]

Reduction in plant parasitic nematode populations was obtained by adding chitin[38] and collagen[39] to the soil. These substrates stimulated soil organisms that could breakdown chitin and collagen. These same organisms were detrimental to nematodes and reduced the soil nematode population that interacts with the nematodes.

VIII. MODIFYING THE FACTORS THAT INVOLVE THE VIRULENCE OF THE PATHOGEN

If genes could be inserted into parasites like bacteria, fungi, and even higher plant parasites that would reduce the virulence, it would be an effective biological control. Organisms with this nonvirulent characteristic could be added to the population in order to reduce the inoculum of virulent pathogen and over time should dilute the inoculum level and provide a type of biological control. If an antisense gene could be inserted into the host which would interfere with the production of the virulent characteristic, then the pathogen would be nonvirulent.

A.   VIRUS SYNTHESIS BLOCKAGE

An effective application of molecular biology to biological control involves putting a gene into the host that interferes with the reproduction of viruses.[40] The classical example is the adding of an antisense gene to the host plant which interferes with the removal of the capsin proteins as the virus enters a host. The net effect is that the antisense gene expression blocks the removal of the capsin protein from the entering virus, and the virus is not able to reproduce.[41] Kunik et al.[42] reported that transgenic tomato plants expressing the capsid protein of the tomato yellow leaf curl virus were resistant to the virus.

Another approach has been to put in a mutated virus replicase gene. The gene interferes with the synthesis of the virus and the net effect is that the transgenic plant is resistant to the virus.[43]

B.   BLOCKAGE OF VIRUS MOVEMENT FROM
CELL TO CELL

Another approach to virus control is to add a gene to the host plant that would prevent the movement of the entering virus from one cell to another through the plasmodesmata. In the case of TMV, a 30-kDa movement protein is produced that interacts with the interconnecting plasmodesmata. When the movement protein gene is added to a transgenic tobacco plant, the exclusion limit is increased from about 800 Da to 9400 Da, which effectively blocked the movement of the TMV virus through the plasmodesmata.[44] Under these conditions, even though the virus enters one cell and is able to reproduce, it would not spread to other cells because of the plasmodesmata blockage between cells. This would effectively control the virus infection.

IX.   SUMMARY

Controlling pests and diseases is a significant factor in the production of food and fiber crops. Many approaches are used today, but biological control is an appealing method. The breeding of crop plants with resistance to pests and pathogens is a successful approach, but is limited in many cases because of the lack of resistant genetic traits. With molecular biology, the techniques are now available for transferring resistant genes from other sources to the crops concerned. The approaches include transferring resistant genes to the host plant, modifying related organisms that compete with the pathogen or pest, and modifying the factors that involve the virulence of the pathogen such as blockage of virus synthesis in the host cell or blockage of virus movement from cell to cell.

REFERENCES

1. Cramer, H. H., *Plant Protection and Crop Production, Pflantzenschutz-Nahr.,* Vol 20, Farbenfabriken Bayr AG, Leverkusen, 1967.
2. Food and Agriculture Organization (FAO), *Production Yearbook, FAO,* Rome, Italy, 1982.
3. Baker, R. R. and Dunn, P. E., Eds., *New Directions in Biological Control, Alternatives for Suppressing Agricultural Pests and Diseases,* Alan R. Liss, New York, 1990.
4. Debach, P. and Rosen, D., *Biological Control by Natural Enemies,* Cambridge University Press, New York, 1991.
5. Burge, M. N., *Fungi in Biological Control Systems,* Manchester University Press, New York, 1988.
6. Goeden, R. D and Kok, L. T., Comments on a proposed 'new' approach for selecting agents for biological control of weeds, *Can. Entomol.,* 118, 51, 1986.
7. Campbell, R., *Biological Control of Microbial Plant Pathogens,* Cambridge University Press, New York, 1989.
8. Misaghi, L. J., *Physiology and Biochemistry of Plant-Pathogen Interactions,* Plenum Press, New York, 1982.
9. Agrios, G. M., *Plant Pathology,* Academic Press, San Diego, CA, 1988.
10. Aronson, A. I., Beckman, W., and Dunn, P., *Bacillus thuringiensis* and related insect pathogens, *Microbiol. Rev.,* 50, 1, 1986.
11. TeBeest, D. O., *Microbial Control of Weeds,* Chapman and Hall, New York, 1991.
12. Davis, L. G., Kuehl, and Battey, J. F., Eds., *Basic Methods in Molecular Biology,* 2nd ed., Appleton and Lange, Norwalk, CT, 1994, 792 pp.
13. Sambrook, J., Fritsch, E. F., and Maniatis, T., *Molecular Cloning: a Laboratory Manual,* 2nd ed., 1,2,3, Cold Spring Harbor, NY, 1989, 1626 pp.
14. Mullis, K. B., Ferre, F., and Gibbs, R. A., Eds., *The Polymerase Chain Reaction,* Springer-Verlag, New York, 1994, 480 pp.
15. Griffin, H. G. and Griffin, A. M., Eds., *PCR Technology, Current Innovations,* CRC Press, Boca Raton, FL, 1994, 400 pp.
16. Tenover, F. C. and Unger, E. R., Nucleic acid probes for detection and identification of infectious agents, in *Diagnostic Molecular Microbiology Principles and Applications,* D. H. Persing, Smith, T. F., Tenover, F. C., and White, T. J., Eds., American Society for Microbiology, Washington D.C., 1993.
17. Andersen, W. R. and Fairbanks, D. J., Molecular markers: important tools for plant genetic resource characterization, *Diversity,* 6, 51, 1990.
18. Williams, J. G. K., Kubelik, A., Livak, K. J., Rafalski, J. A., and Tingey, S., V. DNA polymorphisms amplified by arbitrary primers are useful as genetic markers, *Nucleic Acids Res.,* 18, 6531, 1990.
19. Welsh, J. and McClelland, M., Fingerprinting genomes using PCR with arbitrary primers, *Nucleic Acids Res.,* 18, 7213, 1990.
20. Jones, M. J. and Dunkle, L. D., Analysis of *Cochliobolus carbonum* races by PCR amplification with arbitrary and gene specific primers, *Phytopathology* 83, 366, 1993.
21. Cohn, S. N., Chang, A. C. Y., and Boyer, H. W., Construction of biologically function bacterial plasmids in vitro, *Proc. Natl. Acad. Sci. U.S.A.,* 70, 3240, 1973.
22. Gelvin, S. B., Schilperoort, R. A., and Verma, D. P. S., *Plant Molecular Biology Manual,* Kluwer Academic Publishers, Boston, 1991.
23. Smith, G. E., Summers, M. D., and Fraser, M. J., Production of human beta interferon in insect cells infected with a baculovirus expression vector, *Mol. Cell. Biol.,* 3, 2156, 1983.

24. Chet, I., *Biotechnology in Plant Disease Control*, Wiley-Liss, New York, 1993.
25. Loesch-Fries, L. S., Transgenic plants resistant to viruses, in *New Directions in Biological Control, Alternatives for Suppressing Agricultural Pests and Diseases*, Baker, R. R. and Dunn, P. E., Eds., Alan R. Liss, New York, 1990, 629 pp.
26. Beachy, R. N., Loesch-Fries, S., and Tumer, N. E., Coat protein-mediated resistance against virus infection, *Annu. Rev. Phytopathol.*, 28, 451, 1990.
27. Hoffman, C., Vanderbruggen, H., Hofte, H., Van Rie, J., Jansens, S., and Van Mellaert, H., Specificity of *Bacillus thuringiensis* delta-endotoxin is correlated with the presence of high affinity binding sites in the brush border membrane of target insect midguts, *Proc. Natl. Acad. Sci. U.S.A.*, 85, 7844, 1988.
28. Perlak, F. J., Fuchs, R. L., Dean, D., McPherson, S. L., and Fischhoff, D. A., Modification of the coding sequence enhances plant expression of insect control proteins genes, *Proc. Natl. Acad. Sci. U.S.A.*, 88, 3324, 1991.
29. Powell, P. A., Stark, D. M., Sanders, P. R., and Beachy, R. N., Protection against tobacco mosaic virus in transgenic plants that express tobacco mosaic virus antisense RNA, *Proc. Natl. Acad. Sci. U.S.A.*, 86, 6949, 1989.
30. Neuhaus, A., Ahll-Goy, P., Hinz, U., Flores, S., and Meins, F., High-level expression of a tobacco chitinase gene in *Nicotiana sylvestris*, susceptibility of transgenic plants to *Cercospora nicotianae* infection, *Plant Mol. Biol.*, 16, 141, 1991.
31. Feitelson, J. S., Quick, T. C., and Gaertner, F., Alternative host for *Bacillus thuringiensis* delta-endotoxin genes, in *New Directions in Biological Control, Alternatives for Suppressing Agricultural Pests and Diseases*, Baker, R. R. and Dunn, P. E., Eds., Alan R. Liss, New York, 1990, 561.
32. Maeda, S., Increased insecticidal effect by a recombinant baculovirus carrying a synthetic hormone gene, *Biochem. Biophys. Res. Commun.*, 165, 1177, 1989.
33. Bonning, B. C., Hirst, M., Possee, R. D., and Hammock, B. D., Further development of a recombinant baculovirus insecticide expressing the enzyme juvenile hormone esterase from *Heliothis virescens, Insect Biochem. Mol. Biol.*, 22, 453, 1992.
34. Shapiro, M., Bell, R. A., and Owens, C. D., *In vivo* mass production of gypsy moth nucleopolyhedrosis virus, in *The Gypsy Moth: Research toward Integrated Pest Management,* Doane, C. C. and McManus, M. L., Eds., *For. Serv. Tech. Bull.* 1584, U.S. Department of Agriculture, Washington, D. C., 1981, 633.
35. Martens, J. W. M., Honee, G., Zuidema, D., vanLent, J. W. M., Viss, B., and Viak, J. M., Insecticidal activity of bacterial crystal protein expressed by a recombinant baculovirus in insect cells, *Appl. Environ. Microbiol.*, 56, 2764, 1990.
36. Bishop, D. H. L. and Possee, R. D., Planned release of an engineered baculovirus insecticide, in *New Directions in Biological Control, Alternatives for Suppressing Agricultural Pests and Diseases,* Baker, R. R. and Dunn, P. E., Eds., Alan R. Liss, New York, 1990, 609.
37. Wood, H. A. and Granados, R. R., Genetically engineered baculoviruses as agents for pest control, *Annu. Rev. Microbiol.*, 45, 69, 1991.
38. Spiegel, Y., Cohn, E., and Chet, I., Use of chitin for controlling plant parasitic nematodes: direct effects on nematode reproduction and plant performance, *Plant Soil*, 95, 87, 1986.
39. Galper, S., Cohn, E., Spiegel, Y., and Chet, I., Nematicidal effect of collagen-amended soil and the influence of protease and collagenase. *Rev. Nematol.*, 13, 67, 1990.
40. Cuozzo, M., O'Connell, K. M., Kaniewski, W., Fang, R. X., Chua, N. H., and Tumer, N. E., Viral protection in transgenic tobacco plants expressing the cucumber mosaic virus coat protein and its antisense RNA. *Biotechnology*, 6:549, 1988.

41. Powell, A. P., Nelson, R. S., De, B., Hoffmann, N., Roberts, S. G., Fraley, R. T., and Shad, D. M., Delay of disease development in transgenic plants that express the tobacco mosaic virus coat protein gene, *Science*, 232, 738, 1986.

42. Kunik, T., Salomon, R., Zamir, D., Navot, N., Zeidan, M., Michelson, I., Gafni, Y., and Czosnek, H., Transgenic tomato plants expressing the tomato yellow curl virus capsid protein are resistant to the virus, *Biotechnology*, 12, 500, 1994.

43. Donson, J., Kearny, C. M., Turpen, T. H., Khan, I. A., Kurath, G., Turpen, A. M., Jones, G. E., Dawson, W. O., and Lewandowski, D. J., Broad resistance to tobamoviruses is mediated by a modified tobacco mosaic virus replicase transgene, *Mol. Plant-Microbe Interact.*, 6, 635, 1993.

44. Wolf, S., Deom, C. M., Beachy, R. N., and Lucas, W. L., Movement protein of tobacco virus modifies plasmodesmatal size exclusion limit, *Science*, 246, 337, 1989.

2 Molecular Approaches to Biological Control of Virus Diseases of Plants

Rebecca Grumet and
Tatiana Lanina-Zlatkina

TABLE OF CONTENTS

I. INTRODUCTION

Viruses are obligate pathogens that are entirely dependent upon the host cell for carrying out all metabolic aspects of their life cycle. Since chemical control options are not generally available, viral disease control has largely depended on biologically based control methods including cultural practices (quarantine, sanitation, rotation), cross protection, and genetic resistance.[1] In the past several years, the potential for molecular genetics to contribute to biologically based control methods has been demonstrated for a broad array of plant viruses. This review will focus on the use of viral parasites and the development of pathogen-derived resistance as newly emerging methods to control virus diseases of plants. Other recent reviews related to this subject include those by: Beachy et al.,[2] Fitchen and Beachy,[3] Gadani et al.,[4] Grumet,[5,6] Hull and Davies,[7] Nelson et al.,[8] Scholthof et al.,[9] Tepfer,[10] and Wilson.[11] A list of names and abbreviations of the viruses mentioned in the text is provided in Table 1.

II.   VIRAL PARASITES

The most widely accepted definitions of biological control involve the use of natural enemies such as predators, parasites, or antagonists to control pests.[12,13]

TABLE 1

Names and Abbreviations of Viruses Referred to in the Text

Abbreviation	Name	Virus group
ACMV	African cassava mosaic virus	Geminivirus
AlMV	Alfalfa mosaic virus	Alfalfa mosaic virus group
ArMV	Arabis mosaic virus	Nepovirus
BCTV	Beet curly top virus	Geminivirus
BMV	Brome mosaic virus	Bromovirus
BNYVV	Beet necrotic yellow vein virus	Furovirus
BYMV	Bean yellow mosaic virus	Potyvirus
CLRV	Cherry leafroll virus	Nepovirus
CMV	Cucumber mosaic virus	Cucumovirus
CTV	Citrus tristeza virus	Closterovirus
CyRSV	Cymbidium ringspot virus	Tombusvirus
GCMV	Grapevine chrome mosaic virus	Nepovirus
PEBV	Pea early browning virus	Tobravirus
PLRV	Potato leafroll virus	Luteovirus
PRSV	Papaya ringspot virus	Potyvirus
PSV	Peanut stunt virus	Cucumovirus
PVX	Potato virus X	Potexvirus
PVY	Potato virus Y	Potyvirus
SYNV	Sonchus yellow net virus	Rhabdovirus
TAV	Tomato aspermy virus	Cucumovirus
TGMV	Tomato golden mosaic virus	Geminivirus
TMV	Tobacco mosaic virus	Tobamovirus
TobRV	Tobacco ringspot virus	Nepovirus
TSV	Tobacco streak virus	Ilarvirus
TSWV	Tomato spotted wilt virus	Tospovirus
ZYMV	Zucchini yellow mosaic virus	Potyvirus

Despite their extreme simplicity and complete dependence on the host's cells for all metabolic activity, viruses in several groups are capable of acquiring accompanying nucleic acid sequences that are not essential for viral function and so may be viewed as molecular parasites.[14] These sequences include viral satellites and defective interfering particles. Both types of sequences can have a strong suppressive influence on multiplication of the helper virus and/or symptom severity. Although their role in nature is unknown, it has been suggested that, especially in severe disease cases, selection pressure may favor their maintenance to ensure host survival and virus persistence.[15,16] In recent years, these sequences have become potential tools for the genetic engineering of virus resistance.

A.   VIRAL SATELLITES

Viral satellites are defined as viruses or nucleic acids that depend on a helper virus for replication, are dispensable for the replication of the helper virus, and lack appreciable sequence homology with the helper viral genome; their origins are unknown.[14,17-22] To date, satellites only have been found in association with approximately 5% of plant viruses. Many plant virus satellites are able to interfere with the virus life cycle and suppress symptom expression in the host plant. Despite more than 30 years of study, the molecular mechanisms of satellite interference in the disease development process are still poorly defined. Nevertheless, satellite-based virus control strategies are being successfully developed on an empirical basis in parallel with fundamental satellite research. This has resulted in the exploration of promising technologies for plant protection, including preinoculation with satellite containing viral strains and construction of transgenic satRNA-expressing plants.

The most widely studied family of satRNAs and the one with the most potential benefit for agriculture is associated with CMV, a severe pathogen of nearly 100 crops worldwide.[23] In many cases, the presence of a satRNA in a CMV-infected plant can cause symptom attenuation and a reduction in virus accumulation.[21] In 1982, Jacquemond[24] reported that tomato plants preinoculated with a combination of a mild CMV and a CMV satRNA were protected against severe symptoms when later challenged by a necrogenic CMV strain. Numerous subsequent laboratory studies within the past decade demonstrated similar protection with the use of a variety of CMV and CMV satRNA strains, different host plants, and mechanical as well as aphid (the natural CMV vector) transmission.[25-29] In the cases where it was examined, virus multiplication was suppressed. Comparison of the satellite-free and satellite-containing inoculum showed that the observed protection was due to a combination of classical cross protection (the ability of a mild virus strain to prevent or limit subsequent damage by a more severe strain[30,31]) and the effect of the satRNA itself.[24,26,28,29]

CMV satellites also have been used to limit virus infection in the field. Preinoculation of pepper plants with a combination of an attenuating CMV satRNA and mild CMV strain was tested in China from 1981–1986. Vaccinated fields exhibited an 11–56% increase in yield and a 22–83% decrease in disease index relative to control fields.[25,27,32,33] Similarly, experiments conducted in Italy demonstrated the potential of CMV satRNA to prevent tomato necrosis in epidemic prone regions.[34] In 1988 and 1989, a necrogenic strain of CMV satRNA caused a devastating outbreak in tomato, peppers, and cucurbits. Interspersed throughout a commercial tomato field was a subset of individuals that had been preinoculated with a mild CMV-CMV satRNA combination. Not only were the preinoculated individuals protected, but the field as a whole had a disease index of only 40% relative to the control field.

In the past few years, satellite-mediated virus control has also been achieved by the development of transgenic plants expressing satellite sequences. Potential advantages over the use of preinoculation include reduced labor requirements and the absence of a mild infection resulting from the preinoculating virus. The first examples of engineered satRNA-mediated protection were reported in 1987; transgenic tobacco expressing CMV or TobRV satRNAs were shown to be protected against the severe effects of their respective helper viruses.[35,36] Since then, there have been several other reports of CMV satRNA-mediated resistance in transgenic tomato[37–40] and tobacco.[41–44]

The protected satRNA-expressing transgenic plants were engineered to carry 1–3 monomeric or oligomeric cDNA copies of the satRNA sequences in the plus sense orientation;[35,37,40,42,44] minus-sense satRNA transgenes were less effective in conferring protection.[36,44] In most cases, the transgenic plants showed no abnormalities in growth, flowering, seed production or progeny viability (e.g., see References 37, 40, 42, and 45) nor did they express satellite-characteristic symptoms.[37,38,42] Upon challenge inoculation with the helper virus, infection was established in all cases. Large amounts of unit length satRNAs were synthesized from precursor transcripts in inoculated[36,45] and systemically infected leaves,[35,37–42,45] indicating that the virus was capable of replicating and processing the transgene-derived satRNA transcripts. Regardless of the level of satRNA expression in noninfected transgenic seedlings, similar large amounts of satRNA were produced upon viral infection.[35,41] The amplified satRNA molecules became incorporated into virions of the infecting helper virus and could be transmitted via inocula made from infected sap of the transgenic plants.[41,45] The symptom-modulating properties of the transgene-derived satRNA was maintained in subsequent passages.[41,45]

Transgenic plants expressing satRNAs were significantly protected from the severe symptoms generally associated with the infecting helper virus.[35,36,39–41,43,44] Symptoms exhibited by transgenic plants expressing satRNA resembled those produced in nontransgenic plants inoculated with satRNA containing viral inoculum. Protection has been shown to be effective against aphid transmission[41] and mechanical inoculation with leaf extracts,[35,41] purified virus,[39,43] and viral RNA.[40] The degree of protection was shown to be insensitive to inoculum concentration and independent of the level of the satRNA precursor transcripts expressed prior to inoculation, presumably due to the amplification of the transcripts by the incoming helper virus.[35,41] Symptom attenuation was accompanied by reduced viral RNA, coat protein accumulation, and/or virus titer.[35,36,40,41,44]

A consistent feature of transgenic tomato and tobacco plants expressing CMV satRNA was an absence of noticeable, early protective effects.[35,40,45] Despite the abundance of satRNA transcripts in inoculated leaves, symptoms developed normally in the inoculated and immediately adjacent upper leaves. Consistent with these observations, a delay in the symptom modulating effect of the CMV satRNA was observed in nontransgenic pepper, tobacco, and tomato plants upon pre- or co-inoculation with CMV and CMV satRNA.[25,27,29,46] In contrast, the TobRV satellite appears to confer protection sooner. Even the inoculated leaves of TobRV satRNA-expressing plants had ameliorated symptoms and reduced virus accumulation,[36] and the satRNA of TobRV induced early symptom suppression in inoculated leaves of cowpea.[47] Furthermore, symptom attenuation also was observed with a nonhelper virus (CLRV) even though there was no noticeable replication of the TobRV satRNA.[48] One possible explanation for these results is that CMV satRNA-mediated protection involves a complex series of virus–host interactions, while the TobRV satRNA may influence disease development in a more direct manner.

The extent of specificity of satellite-mediated protection in transgenic plants varies depending on the satRNA strain and/or host plant species. Tomato plants expressing S-CARNA 5 CMV satRNA sequences proved to be disease tolerant when infected by a number of CMV strains, but were not protected against the heterologous cucumovirus, TAV.[40] In contrast, tobacco plants carrying a $I_{17}N$ CMV satRNA transgene were protected against TAV, but not another cucumovirus (PSV) or unrelated viruses.[35]

Interestingly, regardless of whether or not protection was observed against TAV, comparable amounts of unit-length CMV satRNA were accumulated upon infection with either CMV or TAV.[35,40] Thus, as was observed for CMV-inoculated leaves,[35,45] the mere presence of large quantities of satRNA is insufficient for symptom suppression. Furthermore, despite symptom suppression by the $I_{17}N$ CMV satRNA, TAV replication was not affected.[35] Thus, it is also possible to separate the effects on symptom severity from the effects on virus replication. Collectively these observations present further evidence for a complex interaction between satellite, virus, and host components.

B.   DEFECTIVE INTERFERING PARTICLES (DIs)

Defective interfering (DI) particles, as originally defined by Huang and Baltimore,[49] are subviral components composed of partially deleted versions of the parent viral genome. They are encapsidated in a virus-encoded coat protein, require the parent virus for replication, and exhibit the capacity to interfere with replication of the parent virus. The nucleotide sequence of DIs retains large portions of 5′ and 3′ nontranslatable sequences separated by one to several internal nucleotide blocks derived from various coding regions.[50–60] The first clearly characterized DI was found in association with TBSV (a tombusvirus).[54] Since then, a large number of DI RNAs accompanying tombus- and carmoviruses have been extensively characterized.[61] In addition, less studied examples of definite and putative DIs were described within bromo-,[60] furo-,[62–65] gemini-,[51,52,66–69] hordei-,[70] phytoreo-,[53,71] potex-,[57] rhabdo-,[72,73] and tospo-[58,74] virus groups. Many of these have been shown to cause symptom reduction in experimental plants (usually different species of *Nicotiana*) and so may offer an opportunity to control plant virus diseases.[51,52,54,55,59,73–75] The observed symptom attenuation ranges from a delay to complete symptom suppression. Symptom reduction generally is accompanied by a decrease in the level of parent virus;[54,55,72,74–76] an extreme case is the complete elimination of SYNV from uninoculated leaves.[73]

The contribution of DIs to the biology of natural virus populations remains to be determined. At present, the majority of known DIs accompany laboratory rather than field virus isolates, and in some cases the concentration of DI-like components in plant extracts of field virus isolates was much less than typically observed in experimental conditions.[62,64,67] The DI of the furovirus BNYVV presents one instance, however, of symptom attenuation in a natural host.[65] It is quite possible that further investigation will reveal DIs to be a more common component of natural virus populations than is currently recognized.

At this time there are only three examples where DIs have been used for the control of plant virus diseases. *N. benthamiana* plants have been engineered to express DIs of CyRSV, BCTV, and ACMV.[69,77–79] In each case, when inoculated with the parent virus, the transgenic plants amplified the transgene-derived sequence. In comparison with controls, there was a decrease in the accumulation of the parent virus, a delay in symptom development, and a marked reduction in systemic symptom severity. Early symptom expression of ACMV and CyRSV was not reduced. In the case of ACMV, interference with symptom expression and genomic DNA level did not extend to another bipartite geminivirus (TGMV) nor to monopartite BCTV.[77]

As with satRNAs, the genome-derived subgenomic DNA was encapsidated and transmitted along with the parent virus through several mechanical passages.[77] With repeated passaging, however, the relative level of the subgenomic DNA decreased and symptom reduction was less pronounced. On the contrary, the protection effect was enhanced when the virus was transmitted from one transgenic plant to another; in some cases there was a complete arrest of infection.

III. PATHOGEN-DERIVED RESISTANCE

A second approach to the biological control of viruses is cross protection, i.e., protection against infection by a severe viral strain by first inoculating with a mild strain of the virus.[30,31] In this case the protecting strain may be viewed as an antagonist of the severe strain. As is the case for preinoculation with satellites, the use of cross protection is very labor intensive. Commercial applications are generally restricted to high value crops with a long life span, limited hectarage, or small population numbers (e.g., citrus for CTV; papaya for PRSV; and greenhouse tomatoes for TMV).

Despite a variety of studies, including more recent ones utilizing transgenic plants and individual viral genes or gene sequences, the underlying mechanisms of cross protection have not been elucidated. Several hypotheses have been suggested. One hypothesis proposes that the presence of viral coat protein from the first virus interferes with early stages in the life cycle of the incoming, challenge virus (i.e., attachment, entry, or uncoating). Experimental evidence supporting this view includes: observations that viral cross protection frequently can be overcome by inoculating with viral RNA rather than whole virions (e.g., see References 80, 81) and experiments demonstrating that transgenic plants expressing viral coat protein genes can be protected against subsequent viral infection.[2] On the other hand, although viral coat proteins may be involved in cross protection, they are unlikely to be the only factor. Experiments using coat protein mutants,[82,83] transencapsidated viruses,[84] and viroids (small infecting RNA molecules that do not have coat proteins[85]) indicate that the coat protein is not required for cross protection. Investigations with CMV have suggested that for this virus in tobacco at least two mechanisms are involved: one that prevents initial establishment in the challenge inoculated leaf and a second that prevents systemic infection.[86] The host also has been shown to be an important component in the process.[87]

Furthermore, as more types of viral genes and sequences are tested in transgenic plants, it appears that virtually any gene or sequence can be manipulated in a way that will interfere with the establishment of an infection. This is consistent with the concept of pathogen-derived resistance,[88] which states that it should be possible to disrupt the normal pathogenic process by causing a host plant to express a pathogen gene at the wrong time, in the wrong amount, or in a counterfunctional form. Native or altered viral-derived genes potentially can be used to interfere with various stages in the viral life cycle such as uncoating, translation and/or processing, replication, cell to cell or long distance movement, or vector-mediated transmission. Among the viral genes and sequences that have been utilized are coat protein, replicase, movement protein and protease genes, and a variety of antisense and sense defective sequences targeting both coding and noncoding portions of the viral genome.

By the end of 1993, at least 65 examples of engineered plant virus resistance using pathogen-derived genes had been reported in the literature, and the list continues to grow rapidly. There is an increasing number of different viruses, different types of virus genes, and different plant species for which engineered resistance has been achieved. Each type of sequence that has been used successfully to confer resistance will be discussed in turn.

A. COAT PROTEIN GENES

The majority of examples of engineered resistance to date have used viral coat protein (CP) genes. The first demonstration of CP-mediated resistance in plants was for TMV in 1986.[89] Transgenic tobacco plants constitutively expressing the TMV CP gene were more resistant to infection by TMV than were control nontransgenic plants. Since that time, CP-mediated resistance has been accomplished for at least 24 different plant viruses representing at least 13 virus groups that differ in particle morphology and genome composition. Among these are RNA viruses with positive sense, negative sense, single- and double-stranded genomes, and one DNA virus. Resistance is characterized by reduced infection of the inoculated leaf and/or the prevention, delay, or reduction of systemic spread. In the majority of cases, virus accumulation is reduced or absent. As a general rule, CP-mediated resistance is limited to the same virus or closely related viruses and is often overcome by increasing inoculum concentrations. A list of published reports of engineered resistance available as of early 1993 can be found in Grumet.[6]

The mechanism of CP-mediated resistance is not clearly understood. Importantly, experimental results indicate that the mechanism varies among viruses. Perhaps this is not unexpected given the fact that, depending on the virus, CPs can be involved in an array of functions throughout the virus life cycle including encapsidation, cell to cell and long distance movement, vector-mediated transmission, and regulation of replication or translation. One of the earliest stages of viral infection is uncoating of the incoming virus. There is experimental evidence from several systems to show that this can be a stage of interference in CP-expressing plants. Transgenic plants expressing TMV CP and protoplasts derived from those plants were protected against infection by whole virions, but not against infection by viral RNA.[90,91] Similarly, resistance was overcome when the virions were pretreated at pH 8, which causes the particles to swell and increases accessibility of the RNA for uncoating via co-translational disassembly.[91] Other examples where inoculation with RNA overcame coat protein-mediated protection include AlMV[92,93] and TSV.[94]

Interference with uncoating, however, is not likely to be a universal mechanism, nor is it likely to be the only mechanism, even in the examples above. Transgenic plants expressing the coat protein of PVX (potexvirus[95]), PVS (carlavirus[96]), and ArMV and GCMV (nepoviruses[97,98]) were protected against infection even when inoculated by viral RNA. In these cases, some step other than, or in addition to, interference with uncoating must be affected. In the case of CMV, whole plants were protected against systemic infection by both virions and RNA, but protoplasts were protected only against virions and not against RNA.[99] These observations suggest the incoming viral RNA is able to replicate in the primary inoculated cells, but subsequent spread is limited. Okuno et al.[99] suggest that the limited spread may be due to interference with uncoating in adjacent cells, or alternatively, that there may be more than one mechanism operating: one at the cellular level that is likely to involve interference with uncoating, and another at the cell to cell movement or whole plant level. Results with BYMV and TMV also suggest more than one mechanism. Wisniewski et al.[100] found limited spread of TMV after initial infection with RNA, and Hammond and Kamo[101] observed that transgenic plants expressing BYMV CP were initially infected, but leaves that developed later had reduced or no symptoms.

Several of the hypotheses for the mechanism of CP action, such as interference with uncoating, translation, or replication, depend on interaction between the transgene-expressed CP and the viral RNA. There are some instances, however, where the ability to interact with viral RNA may not be sufficient to confer resistance. Tumer et al.[102] showed that changing the second amino acid of the AlMV CP did not alter the ability of the CP to bind AlMV RNA in vitro, but it no longer conferred resistance to infection. Further, although binding and assembly functions of potyviruses are thought to reside in the trypsin-resistant core portion of the CP, transgenic melon plants expressing only the core were only partially resistant to ZYMV infection, whereas those expressing the full length CP were completely resistant.[103] These results suggest that some other sorts of interactions may be involved in the ability of the CPs to confer resistance. The potential interaction with host components is a possibility that has received little attention. Since alterations in coat proteins can influence symptom development and virus spread through the plant, clearly there are points of interaction between the CP and the host that may serve as points of interference in the infection process.

B. REPLICASE GENES

A second and highly promising pathogen-derived resistance approach is the use of viral replicase genes. A key enzyme involved in replication of RNA viruses is a viral encoded RNA-dependent RNA polymerase.[104] These polymerases are characterized by a highly conserved gly-asp-asp (GDD) amino acid motif[104] and also can include a nucleotide (NTP) binding domain.[105] For certain types of plant viruses, the replicase gene is encoded by an open reading frame (ORF) that results in the production of more than one protein; e.g., the replicase ORF of TMV encodes a 126 kDa protein, and a longer, read-through protein of 183 kDa.[106] Within the read-through portion of the 183 kDa gene is the potential for the expression of an additional 54 kDa protein that includes the GDD motif; this putative protein, however, has not been observed *in vivo*.

Golemboski et al.[106] sought to study the function of the putative 54 kDa protein by causing it to be expressed in transgenic plants. Unexpectedly, the transgenic tobacco plants expressing this gene were completely resistant to infection by TMV virions and by RNA at inoculum levels up to 1000-fold higher than for TMV CP-mediated resistance (up to 1 mg/ml). Similarly, certain transgenic tobacco lines expressing the 54 kDa read-through portion of the replicase gene of PEBV were resistant to very high inoculum levels (50 μg/ml to 1 mg/ml PEBV).[107] In both cases the 54 kDa gene transcript was present in the transgenic plants, but there was no detectable 54 kDa protein. However, versions of the 54 kDa genes encoding prematurely truncated proteins did not confer resistance; thus the protein (rather than the RNA) was implicated to be responsible for resistance.[107,108]

Another approach has been to utilize mutated replicase genes. Transformation with a construct encoding a truncated version of the 123 kDa portion of TMV replicase resulted in a high level of protection against both TMV and several other related tobamoviruses;[109] expression of CMV and PVX replicase genes with a deletion or alteration of the GDD motif resulted in resistance to those viruses.[110,111] On the other hand, deletion of the GDD sequence from the replicase gene of PVY did not result in resistance.[112] It is not known if the deletion had other effects which could interfere with ability to confer resistance, such as altering the stability of the protein.

Results from experiments using full length replicase genes have been mixed. In some cases full length replicase genes were used to facilitate infection by replication-defective viruses. Transgenic tobacco plants expressing viral replicase genes of AlMV or BMV could complement inocula that were missing AlMV or BMV replicase components, respectively.[113,114] The replicase gene of the single-stranded geminivirus TGMV complemented a replicase mutant of TGMV.[115] In these cases the replicase gene did not confer resistance against inoculation by replication-competent viruses.

On the other hand, when full length and truncated versions of the replicase genes of PVX[116] and PVY[112] were compared, use of full length genes yielded a higher proportion of resistant lines. Although the extent of resistance varied from minimal to no symptoms for 12 weeks, all lines transformed with the full length replicase gene of PVX showed significantly reduced numbers of lesions relative to controls. Only one resistant line was obtained using the amino terminal half of the PVX replicase gene, and none were obtained using the central portion including the NTP binding domain or the carboxy-terminal portion including the GDD motif. Highly resistant tobacco lines also were produced using the full length replicase gene of CyRSV.[117]

Although the mechanism of replicase-mediated resistance has not been elucidated and, like CP-mediated resistance, may vary among virus groups, distinctive features include a very high level of resistance, expression of resistance at the protoplast level, and protection against infection by RNA as well as virions. These features suggest that the interference is at the replication stage. It is curious that for some viruses, full length replicase genes conferred resistance while for others they facilitated infection. It has been suggested for PVX and PVY that the full length replicases were inhibitory because they either were derived from a virus clone with low infectivity[116] or were somehow altered in the cloning process.[112] An altered replicase gene product might compete with the native polymerase for viral RNA or for host components of the replicase complex. It is also possible, however, that expression of a functional replicase at the wrong time in the life cycle interferes with viral replication by altering the balance between the production of positive and negative RNA strands or by shifting the balance between replication and translation.

C. MOVEMENT PROTEIN AND PROTEASE GENES

Two other types of viral-derived genes that recently have been used to produce virus-resistant transgenic plants are movement protein genes[118,119] and protease genes.[120,121] Successful cell to cell transport of viruses depends on a combination of both host- and virus-encoded factors.[122] For example, TMV encodes a 30 kDa movement protein (MP) that facilitates cell to cell movement by interaction with the interconnecting plasmodesmata. When the TMV MP was expressed in transgenic tobacco plants, the macromolecular exclusion limit of the plasmodesmata was increased from approximately 800 daltons to 9400 daltons.[123] Furthermore, expression of this protein facilitated the movement of a transport-defective TMV strain.[124]

To test the possibility that the MP gene could be used as a resistance gene, Malyshenko et al.[119] and Lapidot et al.,[118] transformed tobacco plants with transport-defective TMV MP genes. There was greatly reduced, or nondetectable virus accumulation in both mature, inoculated leaves, and upper, noninoculated leaves. The mutant TMV-MP also conferred protection against other tobamoviruses.[118] Malyshenko et al.[119] also produced transgenic tobacco plants expressing the native movement protein of BMV, a virus that can replicate in tobacco protoplasts but cannot spread through the plant. Although TMV could replicate in protoplasts prepared from these plants, infection of inoculated and noninoculated leaves was greatly reduced. These examples demonstrate that expression of a defective (or nonspecies appropriate) MP can limit the ability of the virus to spread in the plant.

Another critical step in the life cycle of certain viruses is proteolytic processing. Many viruses express several or all of their genes as part of a polyprotein that is subsequently proteolytically cleaved into individual, mature viral proteins. The potyviruses, for example, express their entire genome as a single polyprotein.[125] Within the viral genome are three cistrons that encode protease activity, one of which, the NIa (nuclear inclusion) protein, is responsible for the majority of the processing. When tobacco plants were transformed with the NIa gene of TVMV,[120] they were resistant to TVMV infection, even at very high inoculum concentrations. Similarly, two of fifty PVY NIa transgenic lines exhibited a strong delay in resistance, or did not develop symptoms.[121] The TVMV NIa-mediated resistance was very specific and did not extend to related potyviruses such as PVY or TEV.[120] Although it is not clear how a non-mutant NIa gene would interfere with the viral infection process, proteases may be an additional type of viral gene that can be used to engineer virus resistance.

D. RNA-MEDIATED RESISTANCE

The resistance strategies described thus far have depended on expression of various viral proteins. There are also cases, however, where viral RNA rather than the encoded protein appears to be the critical factor for conferring resistance. Since viruses are essentially packaged nucleic acids that serve as templates for replication, translation, and in some cases transcription (for DNA viruses), it is not surprising that nucleic acids can be used to interfere with normal viral life cycles. RNA mediated-resistance approaches have utilized both sense (or sense-defective) and antisense RNAs targeting both coding and noncoding portions of the genome.

One approach to RNA-mediated resistance is to introduce genes that encode the complementary or antisense RNA strand. This approach has often been used to down-regulate levels of gene expression in various prokaryotic and eukaryotic systems, presumably by the formation of an RNA duplex that interferes with normal messenger RNA stability, processing, transport, or translation.[126] When dealing with viruses, however, antisense RNAs also might impact processes other than gene expression. They might interfere with viral replication, either through duplex formation or by competition for viral or host factors essential for replication. Similarly, because RNA viruses replicate by synthesizing plus strands from minus and minus strands from plus, sense-defective RNAs (positive RNA strands that have been altered so as to be untranslatable) also might be utilized to interfere with replication.

Several portions of plant viral genomes have been tested for their potential in conferring RNA-mediated resistance. Expression of the antisense sequence of the AL1 gene involved in replication of the TGMV geminivirus led to decreased TGMV infection.[127] Viral noncoding regions have been of particular interest because they include viral regulatory, replication, and translation initiation sites, but the results have been equivocal. After testing several constructs spanning the 5′ and 3′ noncoding

Pershore College

regions of CMV and finding that only one tobacco line from one of the constructs exhibited resistance, Rezian et al.[128] suggested that somaclonal variation could not be ruled out as an explanation for the resistance. In contrast, Nelson et al.[129] were encouraged that several tobacco lines with decreased susceptibility were obtained after transformation with the 5′ nontranslated lead sequence of TMV.

The majority of antisense and sense-defective constructs tested to date have utilized coat protein sequences. In many cases the level of protection conferred by antisense sequences was considerably lower than that conferred by the corresponding coding sequence of the CP genes (e.g., CMV,[130] PVX,[95] TMV,[131] ZYMV,[103] PVY,[132]). In other examples with luteoviruses and potyviruses, however, both the antisense and sense strand RNAs appeared to be comparably effective (BYMV,[101] PLRV,[133]). Similarly, comparable or better levels of protection were observed for plants expressing translation-defective vs. translatable constructs of CP genes of tospoviruses and potyviruses (TSWV,[134] TEV,[135,136] PVY[132,137]). In the cases where it was tested, resistance was limited to viruses with a high degree of sequence homology.[134–136,138] This observation might be expected for a resistance mechanism based on RNA–RNA hybridization.

Intriguing results were reported by Pang et al.,[138] whose work with translatable and nontranslatable versions of the nucleocapsid gene of TSWV suggested that the protein was responsible for conferring protection against heterologous viruses (presumably because heteroencapsidated particles are somewhat dysfunctional), whereas the RNA appeared to be responsible for conferring protection against virus strains with very high sequence homology. Often a lack of correlation between level of protein expression and level of resistance has been one of the factors suggesting RNA-mediated resistance, but Pang et al.[138] also observed a lack of positive correlation between level of RNA expression and level of resistance. Transgenic plants expressing high levels of RNA were less protected than those expressing low levels of RNA.

Lindbo et al.[139] observed a similar phenomenon. Susceptible leaves of transgenic tobacco plants expressing CP genes of TEV had higher transcript levels than did the upper leaves, which did not show symptoms and were not susceptible to reinoculation. They proposed a model suggesting that the transgene transcript and the replicating TEV genome could act additively to trigger a natural cellular response that leads to the inactivation of both the transgene RNAs and incoming viral RNAs. For example, a cytoplasmic, host-encoded RNA-dependent, RNA-polymerase might be stimulated to copy portions of the viral or transgene RNAs and thereby cause the formation of partially double-stranded RNAs that would be targeted for degradation. Although this is a working model with many questions to be addressed, the concept of host-transgene-virus interactions is likely to be important for our understanding of pathogen-derived resistance. Clearly, the viral gene product and incoming virus must interact within the context of the host environment.

V.   FUTURE PROSPECTS

In summary, the use of molecular biological techniques has contributed to our understanding of and ability to develop biological control methods for virus diseases.

Ultimately, the success of these methods will depend on performance in the field. Viral-derived resistance genes now have been used to confer resistance in an array of crops including: tobacco, tomato, potato, cucumber, melon, squash, papaya, alfalfa, rapeseed, rice, and corn. At least one company is nearly ready for commercial production of virus-resistant transgenic crop cultivars (Upjohn, virus-resistant squash, H. Quemada, personal communication). Although relatively few studies of field performance of genetically engineered virus-resistant plants have been published to date, (probably in large part because the majority have been performed by industry) the results have been encouraging.

In the cases where comparisons were reported for the growth and yield of transgenic and control plants (e.g., TMV CP-expressing tomato plants,[140] PVX- and PVY-CP-expressing potato lines[141]), the performance of noninoculated transgenic lines was equal to that of the control nontransgenic lines. Thus, there does not appear to be a negative effect of the introduced gene(s) per se. In those studies and others (CMV CP, cucumber;[142] PVX CP, potato[143]), when the plants were inoculated by virus or allowed to acquire virus by natural vector-mediated transmission, the percentage of infection was significantly less for the transgenics than the controls.

As is the case to some extent for any biological control system that depends on parasites or pathogens,[144] there are potential ecological concerns associated with these virus control strategies.[6,10,30,145–149] Given the ability of transgene-derived satRNA or DI sequences to be encapsidated and transmitted by the helper virus, the high rate of satRNA evolution,[150,151] the high potential for DI recombination,[56,151–155] known minor differences between benign and virulent satellites,[156–158] and the variability of the satRNA or DI phenotype depending on the plant host and the helper virus,[55,60,159–163] the use of viral parasites has been viewed very cautiously.[6,10,11,22,35,45,164] Similar sorts of concerns (that the mild strain itself could cause adverse effects, that some combinations of the mild and severe strains could be detrimental, or that the mild strain could mutate to a more severe strain) also have contributed to the limited use of conventional cross protection strategies.[30]

Among the most frequently mentioned concerns associated with pathogen-derived resistance are altered vector specificity due to transencapsidation of the incoming virus, and genetic recombination between transgene-expressed RNAs and the infecting viral genome. Although several experiments have demonstrated the possibility of transencapsidation (e.g., References 165–168) or recombination (References 169–171), at least among homologous or closely related viruses under conditions of strong selection pressure, the potential ecological impact of these types of events is difficult to discern. Subjects for debate include: the potential for transencapsidation or recombination between viruses that are sufficiently different to have different host ranges or vector specificities, the effect of conditions where recombination is likely to be selectively disadvantageous, and the relative risks associated with transgene-expressed sequences vs. natural mixed infections.

Ultimately, decisions will have to be made that balance the potential benefits, costs, and risks of accepting crop losses due to virus infection, using conventional methods of virus control, or using engineered plants expressing virus-derived sequences. Each option has its own set of advantages and disadvantages that will vary depending on the crop, the virus, and the production system. Although there are still many questions to be answered regarding mechanism, performance, and impact, pathogen-derived resistance and the use of viral parasites are promising strategies for the future reduction of crop losses due to virus infections. In the next several years, we will undoubtedly continue to see a rapidly increasing list of types of genes and different viruses for which these strategies may be applied.

ACKNOWLEDGMENTS

We thank Dr. P. Traynor for critically reviewing the manuscript. T. Lanina-Zlatkina was supported by a grant from the Office of Agriculture, Bureau for Research and Development, U.S. Agency for International Development under Cooperative Agreement No. DAN-4197-A-00-1126-00.

REFERENCES

1. Dunez, J., Perspectives in the control of plant viruses, in *Innovative Approaches to Plant Disease Control*, Chet, I., Ed., Wiley Series in Ecological and Applied Microbiology, John Wiley & Sons, New York, 1987, 297.
2. Beachy, R. N., Loesch-Fries, S., and Tumer, N. E., Coat protein-mediated resistance against virus infection, *Annu. Rev. Phytopathol.*, 28, 451, 1990.
3. Fitchen, J. H. and Beachy, R. N., Genetically engineered protection against viruses in transgenic plants, *Annu. Rev. Microbiol.*, 47, 739, 1993.
4. Gadani, F., Mansky, L. M., Medici, R., Miller, W. A., and Hull, J. H., Genetic engineering of plants for virus resistance, *Arch. Virol.*, 115, 1, 1990.
5. Grumet, R., Genetically engineered plant virus resistance, *HortScience*, 25, 508, 1990.
6. Grumet, R., Development of virus resistant plants via genetic engineering, *Plant Breed. Rev.*, 12, 47, 1994.
7. Hull, R. and Davies, J. W., Approaches to nonconventional control of plant virus diseases, *Crit. Rev. Plant Sci.*, 11, 17, 1992.
8. Nelson, R. S., Powell, P., and Beachy, R. N., Coat protein-mediated protection against virus infection, in *Genetic Engineering of Crop Plants*, Lycett, G. W. and Grierson, D., Eds., Butterworths, Borough Green, Sevenoaks, Kent, U.K., 1990, 13.
9. Scholthof, K-B. G., Scholthof, H. B., and Jackson, A. O., Control of plant virus diseases by pathogen-derived resistance in transgenic plants, *Plant Physiol.*, 102, 7, 1993.
10. Tepfer, M., Viral genes and transgenic plants. What are potential environmental risks?, *Biotechnology*, 11, 1125, 1993.
11. Wilson, T. M. A., Strategies to protect crop plants against viruses, *Proc. Natl. Acad. Sci. U.S.A.*, 90, 3134, 1993.
12. Wilson, F. and Huffaker, C. B., The philosophy, scope and importance of biological control, in *Theory and Practice of Biological Control*, Huffaker, C. B. and Messenger P. S., Eds., Academic Press, New York, 1976, chap.1.
13. Cate, J. R., Biological control of pests and diseases: integrating a diverse heritage, in *New Directions in Biological Control: Alternatives for Suppressing Agricultural Pests and Diseases,* UCLA Symposia on Molecular and Cellular Biology, Vol. 112, Baker, R. R. and Dunn, P. E., Eds., Alan R. Liss, New York, 1990, 23.

14. Francki, R. I. B., Plant virus satellites, *Annu. Rev. Microbiol.*, 39, 151, 1985.
15. Baulcombe, D., Strategies for virus resistance in plants, *Trends Genet.*, 5, 56, 1989.
16. Knorr, D. A., Mullin, R. H., Hearne, P. Q., and Morris, T. J., De novo generation of defective interfering RNAs of tomato bushy stunt virus by high multiplicity passage, *Virology*, 181, 193, 1991.
17. Kassanis, B., Satellitism and related phenomena in plant and animal viruses, *Adv. Virus Res.*, 13, 147, 1968.
18. Kaper, J. M. and Collmer, C. W., Modulation of viral plant diseases by secondary RNA agents, in *RNA Genetics*, Vol. 3, Domingo, E., Holland, J. J., and Ahlquist, P. P., Eds., CRC Press, Boca Raton, FL, 1988, 171.
19. Simon, A. E., Satellite RNAs of plant viruses, *Plant Mol. Biol. Rep.*, 6, 240, 1988.
20. Fritsch, C. and Mayo, M. A., Satellites of plant viruses, in *Plant Viruses*, Vol.1, Mandahar, C., Ed., CRC Press, Boca Raton, FL, 1989, 289.
21. Collmer, C. W. and Howell, S. H., Role of satellite RNA in the expression of symptoms caused by plant viruses, *Annu. Rev. Phytopathol.*, 30, 419, 1992.
22. Roossinck, M. J., Sleat, D., and Palukaitis, P., Satellite RNAs of plant viruses: structures and biological effects, *Microbiol. Rev.*, 265, 1992.
23. Smith, K. M., Cucumber mosaic virus, in *A Textbook of Plant Virus Diseases*, 3rd ed., Longman, London, 1972, 234.
24. Jacquemond, M., Phenomenes d'interferences entre les deux types d'ARN satellite du virus de la mosaique du concombre. Protection des tomates vis a vis la necrose letale, *C. R. Acad. Sci. Paris, III*, 294, 991, 1982.
25. Tien, P. and Chang, X. H., Vaccination of pepper with cucumber mosaic virus isolates attenuated with satellite RNA, in *Proc. 6th Int. Congr. Virol. Sendai*, 1985, 379.
26. Yoshida, K., Goto, T., and Iizuka, N., Attenuated isolates of cucumber mosaic virus produced by satellite RNA and cross-protection between attenuated isolates and virulent ones, *Ann. Phytopath. Soc. Japan*, 51, 238, 1985.
27. Tien, P., Zhang, X., Qiu, B., Qin, B., and Wu, G., Satellite RNA for the control of plant diseases caused by cucumber mosaic virus, *Ann. Appl. Biol.*, 111, 143, 1987.
28. Wu, G., Kang, L., and Tien, P., The effect of satellite RNA on cross-protection among cucumber mosaic virus strains, *Ann. Appl. Biol.*, 114, 489, 1989.
29. Montasser, M. S., Tousignant, E., and Kaper, J. M., Satellite-mediated protection of tomato against cucumber mosaic virus. I. Greenhouse experiments and simulated epidemic conditions in the field, *Plant Dis.*, 75, 86, 1991.
30. Fulton, R. W., Practices and precautions in the use of cross protection for plant virus disease control, *Annu. Rev. Phytopathol.*, 24, 67, 1986.
31. Ponz, F. and Bruening, G., Mechanisms of resistance to plant viruses, *Annu. Rev. Phytopathol.*, 24, 355, 1986.
32. Tien, P. and Chang, X. H., Control of two seed-borne virus diseases in China by the use of protective inoculation, *Seed Sci. Technol.*, 11, 969, 1983.
33. Tien, P. and Gusui, W., Satellite RNA for the biological control of plant disease, *Adv. Virus Res.*, 39, 321, 1991.
34. Gallitelli, D., Vovlas, C., Martelli, G., Montasser, M. S., Tousignant, M. E., and Kaper, J. M., Satellite-mediated protection of tomato against cucumber mosaic virus. II. Field test under natural epidemic conditions in southern Italy, *Plant Dis.*, 75, 93, 1991.
35. Harrison, B. D., Mayo, M. A., and Baulcombe, D. C., Virus resistance in transgenic plants that express cucumber mosaic virus satellite RNA, *Nature*, 328, 799, 1987.
36. Gerlach, W. L., Llewellyn, D., and Haseloff, J., Construction of a plant disease resistance gene from the satellite RNA of tobacco ringspot virus, *Nature*, 328, 802, 1987.

37. McGarvey, P. B., Kaper, J. M., Avila-Rincon, M. J., Pena, L., and Diaz-Ruiz, J. R., Transformed tomato plants express a satellite RNA of cucumber mosaic virus and produce lethal necrosis upon infection with viral RNA, *Biochem. Biophys. Res. Commun.*, 170, 548, 1990.
38. Tousch, D., Jacquemond, M., and Tepfer, M., Transgenic tomato plants expressing a cucumber mosaic virus (CMV) satellite RNA gene: inoculation with CMV induces lethal necrosis, *C. R. Acad. Sci. Paris, III*, 311, 377, 1990.
39. Saito, Y., Komari, T., Masuta, C., Hayashi, Y., Kumashiro, T., and Takanami, Y., Cucumber mosaic virus-tolerant transgenic tomato plants expressing a satellite RNA, *Theor. Appl. Genet.*, 83, 679, 1992.
40. McGarvey, P. B., Montasser, M. S., and Kaper, J. M., Transgenic tomato plants expressing satRNA are tolerant to some strains of cucumber mosaic virus, *J. Am. Soc. Hortic. Sci.*, 119, 642, 1994.
41. Jacquemond, M., Amselem, J., and Tepfer, M., Gene coding for a monomeric form of cucumber mosaic virus satellite RNA confers tolerance to CMV, *Mol. Plant-Microbe Interact.*, 8, 311, 1988.
42. Masuta, C., Komari, T., and Takanami, Y., Expression of cucumber mosaic virus satellite RNA from cDNA copies in transgenic tobacco plants, *Ann. Phytopath. Soc. Japan*, 55, 49, 1989.
43. Yie, Y., Zhao, F., Zhao, S. Z., Liu, Y. Z., Liu, Y. L., and Tien, P., High resistance to cucumber mosaic virus conferred by satellite RNA and coat protein in transgenic commercial tobacco cultivar G-140, *Mol. Plant-Microbe Interact.*, 5, 460, 1992.
44. Tousch, D., Jacquemond, M., and Tepfer, M., Replication of cucumber mosaic virus satellite RNA from negative-sense transcripts produced either in vitro or in transgenic plants, *J. Gen. Virol.*, 75, 1009, 1994.
45. Baulcombe, D. C., Saunders, G. R., Bevan, M. W., Mayo, M. A., and Harrison, B. D., Expression of biologically active viral satellite RNA from the nuclear genome of transformed plants, *Nature*, 321, 446, 1986.
46. Devic, M., Jaegle, M., and Baulcombe, D., Symptom production in tobacco and tomato is determined by two distinct domains of the satellite RNA of cucumber mosaic virus (strain Y), *J. Gen. Virol.*, 70, 2765, 1989.
47. Schneider, I. R., Characteristics of a satellite-like virus of tobacco ringspot virus, *Virology*, 45, 108, 1971.
48. Ponz, F., Rowhani, A., Mircetich, S. M., and Bruening, G., Cherry leafroll virus infections are affected by a satellite RNA that the virus does not support, *Virology*, 160, 183, 1987.
49. Huang, A. S. and Baltimore, D., Defective viral particles and viral disease process, *Nature*, 226, 325, 1970.
50. Bouzoubaa, S., Guilley, H., Jonard, G., Richards, K., and Putz, C., Nucleotide sequence analysis of RNA-3 and RNA-4 of beet necrotic yellow vein virus, isolates F2 and G1, *J. Gen. Virol.*, 66, 1553, 1985.
51. Stanley, J. and Townsend, R., Characterization of DNA forms associated with cassava latent virus infection, *Nucleic Acids Res.*, 13, 2189, 1985.
52. MacDowell, S. W., Coutts, R. H. A., and Buck, K. W., Molecular characterization of subgenomic single-stranded and double-stranded DNA forms isolated from plants infected with tomato golden mosaic virus, *Nucleic Acids Res.*, 14, 7967, 1986.
53. Anzola, J. V., Zhengkai, X., Asamizu, T., and Nuss, D. L., Segment-specific inverted repeats found adjacent to conserved terminal sequences in wound tumor virus genome and defective interfering RNAs, *Proc. Natl. Acad. Sci. U.S.A.*, 84, 8301, 1987.

54. Hillman, B. I., Carrington, J. C., and Morris, T. J., A defective interfering RNA that contains a mosaic of a plant virus genome, *Cell*, 51, 427, 1987.

55. Li, X. H., Heaton, L. A., Morris, T. J., and Simon, A. E., Turnip crinkle virus defective interfering RNAs intensify viral symptoms and are generated de novo, *Proc. Natl. Acad. Sci. U.S.A.*, 86, 9173, 1989.

56. Rubino, L., Burgyan, J., Grieco, F., and Russo, M., Sequence analysis of cymbidium ringspot virus satellite and defective interfering RNAs, *J. Gen. Virol.*, 71, 1655, 1990.

57. White, K. A., Bancroft, J. B., and Mackie, G. A., Defective RNAs of clover yellow mosaic virus encode nonstructural/coat protein fusion products, *Virology*, 183, 479, 1991.

58. Resende, R. de O., de Haan, P., van de Vossen, E., de Avila, A. C., Goldbach, R., and Peters, D., Defective interfering L RNA segments of tomato spotted wilt virus retain both virus genome termini and have extensive internal deletions, *J. Gen. Virol.*, 73, 2509, 1992.

59. Finnen, R. L. and Rochon, D. M., Sequence and structure of defective interfering RNAs associated with cucumber necrosis virus infections, *J. Gen. Virol.*, 74, 1715, 1993.

60. Romero, J., Huang, Q., Pogany, J., and Bujarski, J. J., Characterization of defective interfering RNA components that increase symptom severity of broad bean mottle virus infections, *Virology*, 194, 576, 1993.

61. Roux, L., Simon, A. E., and Holland, J. J., Effects of defective interfering viruses on virus replication and pathogenesis in vitro and in vivo, *Adv. Virus Res.*, 40, 181, 1991.

62. Shirako, Y. and Brakke, M. K., Spontaneous deletion mutation of soil-borne wheat mosaic virus RNA II, *J. Gen. Virol.*, 65, 855, 1984.

63. Shirako, Y. and Brakke, M. K., Two purified RNAs of soil-borne wheat mosaic virus are needed for infection, *J. Gen. Virol.*, 65, 119, 1984.

64. Burgermeister, W., Koenig, R., Weich, H., Sebald, W., and Lesemann, D-E., Diversity of the RNAs in thirteen isolates of beet necrotic yellow vein virus in *Chenopodium quinoa* detected by means of cloned cDNA, *J. Phytopathol.*, 115, 229, 1986.

65. Koenig, R. and Burgermeister, W., Mechanical inoculation of sugarbeet roots with isolates of beet necrotic yellow vein virus having different RNA composition, *J. Phytopathol.*, 124, 249, 1989.

66. Hamilton, W. D. O., Bisaro, D. M., Coutts, R. H. A., and Buck, K. W., Demonstration of the bipartite nature of the genome of a single-stranded DNA plant virus by infection with the cloned DNA components, *Nucleic Acids Res.*, 11, 7387, 1983.

67. Coutts, R. H. A. and Buck, K. W., Cassava latent virus specific DNAs in mosaic diseased cassava of Nigerian origin, *Neth. J. Plant Pathol.*, 93, 241, 1987.

68. Roberts, E. J. F., Buck, K. W., and Coutts, R. H. A., Characterization of potato yellow mosaic virus as a geminivirus with a bipartite genome, *Intervirology*, 29, 162, 1988.

69. Frischmuth, T. and Stanley, J., Beet curly top virus symptom amelioration in *Nicotiana benthamiana* transformed with a naturally occurring viral subgenomic DNA, *Virology*, 200, 826, 1994.

70. Jackson, A. O., Hunter, B. G., and Gustafson, G. D., Hordeivirus relationships and genome organization, *Annu. Rev. Phytopathol.*, 27, 95, 1989.

71. Reddy, D. V. R. and Black, L. M., Deletion mutations of the genome segments of the wound tumor virus, *Virology*, 61, 458, 1974.

72. Adam, G., Gaedigk, K., and Mundry, K. W., Alterations of a plant rhabdovirus during successive mechanical transfers, *Z. Pflanzenkr. Pflanzenschutz*, 90, 28, 1983.

73. Ismail, I. D. and Milner, J. J., Isolation of defective interfering particles of sonchus yellow net virus from chronically infected plants, *J. Gen. Virol.*, 69, 999, 1988.

74. Resende, R. de O., de Haan, P., de Avila, A. C., Kitajima, E. W., Kormelink, R., Goldbach, R., and Peters, D., Generation of envelope and defective interfering RNA mutants of tomato spotted wilt virus by mechanical passage, *J. Gen. Virol.*, 72, 2375, 1991.

75. Burgyan, J., Grieco, F., and Russo, M., A defective interfering RNA molecule in cymbidium ringspot virus infections, *J. Gen. Virol.*, 70, 235, 1989.

76. Rochon, D. M. and Johnston, J. C., Infectious transcripts from cucumber necrosis virus cDNA: evidence for a bifunctional subgenomic RNA, *Virology*, 181, 656, 1991.

77. Stanley, J., Frischmuth, T., and Ellwood, S., Defective viral DNA ameliorates symptoms of geminivirus infection in transgenic plants, *Proc. Natl. Acad. Sci. U.S.A.*, 87, 6291, 1990.

78. Frischmuth, T. and Stanley, J., African cassava mosaic virus DI DNA interferes with the replication of both genomic components, *Virology*, 183, 539, 1991.

79. Kollar, A., Dalmay, T., and Burgyan, J., Defective interfering RNA-mediated resistance against cymbidium ringspot virus in transgenic plants, *Virology*, 193, 313, 1993.

80. Sherwood, J. L. and Fulton, R. W., The specific involvement of coat protein in tobacco mosaic virus cross protection, *Virology*, 119, 150, 1982.

81. Dodds, J. A., Lee S. Q., and Tiffany, M., Cross protection between strains of cucumber mosaic virus: effect of host and type of inoculum on accumulation of virions and double-stranded RNA of the challenge strain, *Virology*, 144, 301, 1985.

82. Zaitlin, M., Viral cross protection: more understanding is needed, *Phytopathology*, 66, 382, 1976.

83. Gerber M. and Sarkar S., The coat protein of tobacco mosaic virus does not play a significant role for cross protection, *J. Phytopathol.*, 124, 323, 1989.

84. Zinnen, T. M. and Fulton, R. W., Cross-protection between sunn-hemp mosaic and tobacco mosaic viruses, *J. Gen. Virol.*, 67, 1679, 1986.

85. Niblett, C. L., Dickson, E., Fernow, K. H., Horst, R. K., and Zaitlin, M., Cross protection among four viroids, *Virology*, 91, 198, 1978.

86. Dodds, J. A., Approaches to studying viral cross protection, in *New Directions in Biological Control: Alternatives for Suppressing Agricultural Pests and Diseases*, UCLA Symposia on Molecular and Cellular Biology, Vol. 112, Baker, R. R. and Dunn, P. E., Eds., Alan R. Liss, New York, 1990, 23.

87. Rezende, J. A. M., Urban, L., Sherwood, J. L., and Melcher, U., Host effect of cross protection between two strains of tobacco mosaic virus, *J. Phytopathol.*, 136, 147, 1992.

88. Sanford, J. C. and Johnston, S. A., The concept of parasite derived resistance — Deriving resistance genes from the parasite's own genome, *J. Theor. Biol.*, 113, 395, 1985.

89. Powell-Abel, P., Nelson, A. S., Hoffmann, B., De, N., Rogers, S. G., Fraley, R. T., and Beachy, R. N., Delay of disease development in transgenic plants that express the tobacco mosaic virus coat protein gene, *Science*, 232, 738, 1986.

90. Nelson, R. S., Powell-Abel, P., and Beachy, R. N., Lesions and virus accumulation in inoculated transgenic tobacco plants expressing the coat protein gene of tobacco mosaic virus, *Virology*, 158, 126, 1987.

91. Register, J. C. and Beachy, R. N., Resistance to TMV transgenic plants results from interference with an early event in infection, *Virology*, 166, 524, 1988.

92. Loesch-Fries, L. S., Merlo, D., Zinnen, T., Burhop, L., Hill, K., Krahn, K., Jarvis, N., Nelson S., and Halk, E., Expression of alfalfa mosaic virus RNA 4 in transgenic plants confers virus resistance, *EMBO J.*, 6, 1845, 1987.

93. Van Dun, C. M. P., Bol, J. F., and Van Vloten-Doting, L., Expression of alfalfa mosaic virus and tobacco rattle virus coat protein genes in transgenic tobacco plants, *Virology*, 159, 299, 1987.

94. Van Dun, C. M. P., Overduin, B., Van Vloten-Doting, L., and Bol, J. F., Transgenic tobacco expressing tobacco streak virus or mutated alfalfa mosaic virus coat protein does not cross protect against alfalfa mosaic virus infection, *Virology*, 164, 383, 1988.

95. Hemenway, C., Fang, R.-X., Kaniewski, W. K., Chua, N.-H., and Tumer, N. E., Analysis of the mechanism of protection in transgenic plants expressing the potato virus X coat protein or its antisense RNA *EMBO J.*, 7, 1273, 1988.

96. MacKenzie, D. J. and Tremaine, J. H., Transgenic *Nicotiana debneyii* expressing viral coat protein are resistant to potato virus S infection, *J. Gen. Virol.*, 71, 2167, 1990.

97. Bertioli, D. J., Cooper, J. I., Edwards, M. L., and Hawes, W. S., Arabis mosaic nepovirus coat protein in transgenic tobacco lessens disease severity and virus replication, *Ann. Appl. Biol.*, 120, 47, 1992.

98. Brault, V., Candresse, T., le Gall, O., Delbos, R. P., Lanneau, M., and Dunez, J., Genetically engineered resistance against grapevine chrome mosaic nepovirus, *Plant Mol. Biol.*, 21, 89, 1993.

99. Okuno, T., Nakayama, M., Yoshia, S., Furasawa, I., and Komiya, T., Comparative susceptibility of transgenic tobacco plants and protoplasts expressing the coat protein gene of cucumber mosaic virus to infection with virions and RNA, *Phytopathology*, 83, 542, 1993.

100. Wisniewski, L. A., Powell, P. A., Nelson, R. S., and Beachy, R. N., Local and systemic spread of tobacco mosaic virus in transgenic tobacco, *Plant Cell*, 2, 559, 1992.

101. Hammond, J. and Kamo, K. K., Transgenic coat protein and antisense RNA resistance to bean yellow mosaic potyvirus, *Acta Hort.*, 336, 171, 1993.

102. Tumer, N. E., Kaniewski, W., Haley, L., Gehrke, L., Lodge, J. K., and Sanders, P., The second amino acid of alfalfa mosaic virus coat protein is critical for coat protein-mediated protection, *Proc. Natl. Acad. Sci. U.S.A.*, 88, 2331, 1991.

103. Fang, G. and Grumet, R., Genetic engineering of potyvirus resistance using constructs derived from the zucchini yellow mosaic virus coat protein gene, *Mol. Plant-Microbe Interact.*, 6, 358, 1993.

104. Kamer, G. and Argos, P., Primary structural comparisons of RNA-dependent RNA polymerases from plant, animal and bacterial viruses, *Nucleic Acids Res.*, 12, 7269, 1984.

105. Habili, N. and Symons, R. H., Evolutionary relationship between luteoviruses and other RNA plant viruses based on sequence motifs in their putative RNA polymerase and nucleic acid helicases, *Nucleic Acids Res.*, 17, 9543, 1989.

106. Golemboski, D. B., Lomonossoff, G. P., and Zaitlin, M., Plants transformed with a tobacco mosaic virus nonstructural gene sequence are resistant to the virus, *Proc. Natl. Acad. Sci. U.S.A.*, 87, 6311, 1990.

107. MacFarlane, S. A. and Davies, J. W., Plants transformed with a region of the 201-kilodalton replicase gene from pea early browning virus RNA1 are resistant to virus infection, *Proc. Natl. Acad. Sci. U.S.A.*, 89, 5829, 1992.

108. Carr, J. P., Marsh, L. E., Lomonossoff, G. P., Sekiya, M. E., and Zaitlin, M., Resistance to tobacco mosaic virus induced by the 54-kDa gene sequence requires expression of the 54-kDa protein, *Mol. Plant-Microbe Interact.*, 5, 397, 1992.

109. Donson, J., Kearny, C. M., Turpen, T. H., Khan, I. A., Kurath, G., Turpen, A. M., Jones, G. E., Dawson, W. O., and Lewandowski, D. J., Broad resistance to tobamoviruses is mediated by a modified tobacco mosaic virus replicase transgene, *Mol. Plant-Microbe Interact.*, 6, 635, 1993.

110. Anderson, J. M., Palukaitis, P., and Zaitlin, M., A defective replicase gene induces resistance to cucumber mosaic virus in transgenic tobacco plants, *Proc. Natl. Acad. Sci. U.S.A.*, 89, 8759, 1992.

111. Longstaff, M., Brigneti, G., Boccard, F., Chapman, S., and Baulcombe, D., Extreme resistance to potato virus X infection in plants expressing a modified component of the putative viral replicase, *EMBO J.*, 12, 379, 1993.
112. Audy, P., Palukaitis, P., Slack, S. A., and Zaitlin, M., Replicase-mediated resistance to potato virus Y in transgenic tobacco plants, *Mol. Plant-Microbe Interact.*, 7, 15, 1994.
113. Van Dun, C. M. P., Van Vloten-Doting, L., and Bol, J. F., Expression of alfalfa mosaic virus cDNA 1 and 2 in transgenic tobacco plants, *Virology*, 163, 572, 1988.
114. Mori, M., Mise, K., Okuno, T., and Furusawa, I., Expression of brome mosaic virus-encoded replicase genes in transgenic tobacco plants, *J. Gen. Virol.*, 73, 169, 1992.
115. Hanley-Bowdoin, L., Elmer, J. S., and Rogers, S. G., Expression of functional replication protein from tomato golden mosaic virus in transgenic tobacco plants, *Proc. Natl. Acad. Sci. U.S.A.*, 87, 1446, 1990.
116. Braun, C. J. and Hemenway, C. L., Expression of amino-terminal portions or full-length viral replicase genes in transgenic plants confers resistance to potato virus X infection, *Plant Cell*, 4, 735, 1992.
117. Rubino, L., Lupo, R., and Russo, M., Resistance to cymbidium ringspot tombusvirus infection in transgenic *Nicotiana benthamiana* plants expressing a full-length viral replicase gene, *Mol. Plant-Microbe Interact.* 6, 729, 1993.
118. Lapidot, M., Gafny, R., Ding, B., Wolf, S., Lucas, W.J., and Beachy, R.N., A dysfunctional movement protein of tobacco mosaic virus that partially modifies the plasmodesmata and limits virus spread in transgenic plants, *Plant J.*, 4, 959, 1993.
119. Malyshenko, S. I., Kondakova, O. A., Nazarova, J. V., Kaplan, I. B., Taliansky, M. E., and Atabekov, J. G., Reduction of tobacco mosic virus accumulation in transgenic plants producing non-functional viral transport proteins, *J. Gen. Virol.*, 74, 1149, 1993.
120. Maiti, I. B., Murphy, J. F., Shaw, F. G., and Hunt, A. G., Plants that express a potyvirus proteinase gene are resistant to virus infection, *Proc. Nat. Acad. Sci. U.S.A.*, 90, 6110, 1993.
121. Vardi, E., Sela, I., Edelbaum, O., Livneh, O., Kuzentsova, L., and Stram, Y., Plants transformed with a cistron of a potato virus Y protease (NIa) are resistant to virus infection, *Proc. Natl. Acad. Sci. U.S.A.*, 90, 7513, 1993.
122. Deom, C. M., Lapidot, M., and Beachy, R. N., Plant virus movement proteins, *Cell*, 69, 221, 1992.
123. Wolf, S., Deom, C. M., Beachy, R. N., and Lucas, W. L., Movement protein of tobacco mosaic virus modifies plasmodesmatal size exclusion limit, *Science*, 246, 337, 1989.
124. Deom, C. M., Oliver, M. I., and Beachy, R. N., The 30 kilodalton gene product of tobacco mosaic virus potentiates virus movement, *Science*, 237, 389, 1987.
125. Dougherty, W. G. and Carrington, J. C., Expression and function of potyviral gene products, *Annu. Rev. Phytopathol.*, 26, 123, 1988.
126. Eguchi, Y., Ithoh, T. and Tomizawa, J. I., Antisense RNA, *Annu. Rev. Biochem.*, 60, 631, 1991.
127. Day, A. G., Bejarano, E. R., Buck, K. W., Burrell, M., and Lichtenstein, C. P., Expression of an antisense viral gene in transgenic tobacco confers resistance to the DNA virus tomato golden mosaic virus, *Proc. Natl. Acad. Sci. U.S.A.*, 88, 6721, 1991.
128. Rezaian, M. A., Skene, K. G. M., and Ellis, J. G., Anti-sense RNAs of cucumber mosaic virus in transgenic plants assessed for control of the virus, *Plant Mol. Biol.*, 11, 463, 1988.
129. Nelson, A., Roth, D. A., and Johnson, J. D., Tobacco mosaic virus infection of transgenic *Nicotiana tabacum* plants is inhibited by antisense constructs directed at the 5′ region of viral RNA, *Gene*, 127, 227, 1993.

130. Cuozzo, M., O'Connell, K. M., Kaniewski, W., Fang, R.-X., Chua, N.-H., and Tumer, N. E., Viral protection in transgenic tobacco plants expressing the cucumber mosaic virus coat protein or its antisense RNA, *Biotechnology*, 6, 549, 1988.

131. Powell, P. A., Stark, D. M., Sanders, P. R., and Beachy, R. N., Protection against tobacco mosaic virus in transgenic plants that express tobacco mosaic virus antisense RNA, *Proc. Natl. Acad. Sci. U.S.A.*, 86, 6949, 1989.

132. Farinelli, L. and Malnoe, P., Coat-protein gene-mediated resistance to potato virus Y in tobacco: examination of the resistance mechanisms — is the transgenic coat protein required for protection?, *Mol. Plant-Microbe Interact.*, 6, 284, 1993.

133. Kawchuk, L. M., Martin, R. R., and McPherson, J., Sense and antisense RNA-mediated resistance to potato leafroll virus in Russet Burbank potato plants, *Mol. Plant-Microbe Interact.*, 4, 247, 1991.

134. de Haan, P., Gielen, J. J. L., Prins, M., Wijkamp, I. G., van Schepen, A., Peters, D., van Grinsven, M. Q. J. M., and Goldbach, R., Characterization of RNA-mediated resistance to tomato spotted wilt virus in transgenic tobacco plants, *Biotechnology*, 10, 1133, 1992.

135. Lindbo, J. A. and Dougherty, W. G., Pathogen-derived resistance to a potyvirus: immune and resistant phenotypes in transgenic tobacco expressing altered forms of a potyvirus coat protein nucleotide sequence, *Mol. Plant-Microbe Interact.*, 5, 144, 1992.

136. Lindbo, J. A. and Dougherty, W. G., Untranslatable transcripts of the tobacco etch virus coat protein gene sequence can interfere with tobacco etch virus replication in transgenic plants and protoplasts, *Virology*, 189, 725, 1992.

137. van der Vlugt, R. A. A., Ruiter, R. K., and Goldbach, R., Evidence for sense RNA-mediated protection to PVY[N] in tobacco plants transformed with the viral coat protein cistron, *Plant Mol. Biol.*, 20, 631, 1992.

138. Pang, S. Z., Slightom, J. L., and Gonsalves, D., Different mechanisms protect transgenic tobacco against tomato spotted wilt and impatiens necrotic spot tospoviruses, *Biotechnology*, 11, 819, 1993.

139. Lindbo, J. A., Silva-Rosales, J., Proebsting, W. M., and Dougherty, W.G., Induction of a highly specific antiviral state in transgenic plants: implications for regulation of gene expression and virus resistance, *Plant Cell*, 5, 1749, 1993.

140. Sanders, P. R., Sammons, B., Kaniewski, W., Haley, L., Layton, J., LaVallee, B. J., Delannay, X., and Tumer, N. E., Field resistance of transgenic tomatoes expressing the tobacco mosaic virus or tomato mosaic virus coat protein genes, *Phytopathology*, 82, 683, 1992.

141. Kaniewski, W., Lawson, C., Sammons, B., Haley, L., Hart, J., Delannay, X., and Tumer, N. E., Field resistance of transgenic Russet Burbank potato to effects of infection by potato virus X and potato virus Y, *Biotechnology*, 8, 750, 1990.

142. Gonsalves, D., Chee, P., Provvidenti, R., Seem, R., and Slightom, J. L., Comparison of coat protein-mediated and genetically-derived resistance in cucumbers to infection by cucumber mosaic virus under field conditions with natural challenge inoculations by vectors, *Biotechnology*, 10, 1562, 1992.

143. Jongedijk, E., de Schutter, A. A. J. M., Stolte, T., van den Elzen, P. J. M., and Cornelissen, B. J. C., Increased resistance to potato virus X and preservation of cultivar properties in transgenic potato under field conditions, *Biotechnology*, 10, 422, 1992.

144. Pimentel, D., Using genetic engineering for biological control: reducing ecological risks, in *Engineered Organisms in the Environment: Scientific Issues*, Halvorson, H. O., Pramer, D., and Rogul, M., Eds., American Society for Microbiology, Washington, DC, 1985, 129.

145. de Zoeten, G. A., Risk assessment: Do we let history repeat itself?, *Phytopathology* 81, 585, 1991.

146. Falk, B. W. and Bruening, G., Will transgenic crops generate new viruses and new diseases?, *Science*, 264, 1395, 1994.

147. Bruening, G. and Falk, B. W., Risks in using transgenic plants?, *Science*, 264, 1651, 1994.

148. Gibbs, M., Risks in using transgenic plants?, *Science*, 264, 1650, 1994.

149. Hull, R., Risks in using transgenic plants?, *Science*, 264, 1649, 1994.

150. Kurath, G. and Palukaitis, P., RNA sequence heterogeneity in natural populations of three satellite RNAs of cucumber mosaic virus, *Virology*, 173, 231, 1989.

151. Kurath, G. and Palukaitis, P., Serial passage of infectious transcripts of a cucumber mosaic virus satellite RNA results in sequence heterogeneity, *Virology*, 176, 8, 1990.

152. Simon, A. E. and Howell, S. H., The virulent satellite RNA of turnip crinkle virus has a major domain homologous to the 3′ end of the helper virus genome, *EMBO J.*, 5, 3423, 1986.

153. Cascone, P. J., Carpenter, C. D., Li, X. H., and Simon, A. E., Recombination between satellite RNAs of turnip crinkle virus, *EMBO J.*, 9, 1709, 1990.

154. Zhang, C., Cascone, P. J., and Simon, A. E., Recombination between satellite and genomic RNAs of turnip crinkle virus, *Virology*, 184, 791, 1991.

155. White, K. A. and Morris, T. J., Recombination between defective tombusvirus RNAs generates functional hybrid genomes, *Proc. Natl. Acad. Sci. U.S.A.*, 91, 3642, 1994.

156. Masuta, C. and Takanami, Y., Determination of sequence and structural requirements for pathogenicity of a cucumber mosaic virus satellite RNA (Y-satRNA), *Plant Cell*, 1, 1165, 1989.

157. Sleat, D. E. and Palukaitis, P., Site-directed mutagenesis of a plant viral satellite RNA changes its phenotype from ameliorative to necrogenic, *Proc. Natl. Acad. Sci. U.S.A.*, 87, 2946, 1990.

158. Zhang, L., Kim, C. H., and Palukaitis, P., The chlorosis-induction domain of the satellite RNA of cucumber mosaic virus: identifying sequences that affect accumulation and the degree of chlorosis, *Mol. Plant-Microbe Interact.*, 7, 208, 1994.

159. Takanami, Y., A striking change in symptoms on cucumber mosaic virus-infected tobacco plants induced by a satellite RNA, *Virology*, 109, 120, 1981.

160. Waterworth, H. E., Kaper, J. M., and Tousignant, M. E., CARNA 5, the small cucumber mosaic virus-dependent replicating RNA, regulates disease expression, *Science*, 204, 845, 1985.

161. Palukaitis, P., Pathogenicity regulation by satellite RNAs of cucumber mosaic virus: minor nucleotide sequence changes alter host responses, *Mol. Plant-Microbe Interact.*, 1, 175, 1988.

162. Sleat, D. E. and Palukaitis, P., Induction of tobacco chlorosis by certain cucumber mosaic virus satellite RNAs is specific to subgroup II helper strains, *Virology*, 176, 292, 1990.

163. Sleat, D. E., Zhang, L., and Palukaitis, P., Mapping determinants within cucumber mosaic virus and its satellite RNA for the induction of necrosis in tomato plants, *Mol. Plant-Microbe Interact.*, 7, 189, 1994.

164. Courtice, G., Satellite defenses for plants, *Nature*, 328, 758, 1987.

165. Osbourn, J. K., Sarkar, S., and Wilson, T. M. A., Complementation of coat protein-defective TMV mutants in transgenic tobacco plants expressing TMV coat protein, *Virology*, 179, 921, 1990.

166. Holt, C. A. and Beachy, R. N., In vivo complementation of infectious transcripts from mutant tobacco mosaic virus cDNAs in transgenic plants, *Virology*, 181, 109, 1991.

167. Farinelli, L., Malnoe, P., and Collet, G. F., Heterologous encapsidation of potato virus Y strain 0 (PVY^O) with the transgenic coat protein of PVY strain N (PVY^N) in *Solanum tuberosum* cv. Bintje, *Biotechnology*, 10, 1020, 1992.

168. Lecoq, H., Ravelonandro, M., Wipf-Scheibel, C., Monsion, M., Raccah, B., and Dunez, J., Aphid transmission of a non-transmissible strain of zucchini yellow mosaic potyvirus from transgenic plants expressing the capsid protein of plum pox potyvirus, *Mol. Plant-Microbe Interact.*, 6, 403, 1993.

169. Bujarski, J. J. and Kaesberg, P., Genetic recombination between RNA components of a multipartite plant virus, *Nature*, 321, 528, 1986.

170. Rao, A. L. N. and Hall, T. C., Recombination and polymerase error facilitate restoration of infectivity in brome mosaic virus, *J. Virol.*, 67, 969, 1993.

171. Allison, R., Thompson, C., and Ahlquist, P., Regeneration of a functional RNA virus genome by recombination between deletion mutants and requirement for cowpea chlorotic mottle virus 3a and coat genes for systemic infection, *Proc. Natl. Acad. Sci. U.S.A.*, 87, 1820, 1990.

172. Greene, A. E. and Allison, R. F., Recombination between viral RNA and transgenic plant transcripts, *Science*, 263, 1423, 1994.

3 Molecular Biology of the Biological Control of Plant Bacterial Diseases

Alice L. Pilgeram and David C. Sands

TABLE OF CONTENTS

0-8493-2442-4/96/$0.00+$.50

I.   INTRODUCTION

All microorganisms occupy various ecological niches in nature, often on a first-come first-served basis. In order for a plant disease to develop, a pathogen must compete with other microorganisms to secure the necessary niche on the susceptible plant. Competitive activity is especially important to the development of plant disease in cases where the pathogen goes through an epiphytic growth mode prior to attaining infection threshold levels. An effective biological control agent excludes or displaces a pathogen from an available niche by preemption or by direct competition through differential depletion of nutrients or suppression of pathogen growth through the production of antibiotics or bacteriocins.

Biological control of bacterial plant diseases in nature seems to be the rule rather than the exception. In fact, early in this century biological control was the only means to manage plant diseases. During the past 40 years, biological control has been replaced with chemical control in many cultivated crops because of the wide spectrum of chemicals available on the market, the ease of use, and the efficacy of pest control. Increased environmental awareness has recently mandated a switch from chemical controls back to biological alternatives. Unfortunately, implementation of biological control is currently hindered by the lack of effective biological control agents and strategies.

It is clear that wild-type agents taken from nature are seldom adequate for biological control. Molecular techniques allow modification of these potential biocontrol agents by the directed removal or addition of one or more genes. For example, the antagonistic properties of a biocontrol organism towards a target pathogen could be increased by expression of novel antimicrobials.

Biocontrol strategies frequently fail in the field because the biocontrol agent is unable to compete with the pathogen under conditions which favor proliferation of the pathogen and development of disease. One recent approach to this problem is to minimize differences in the biocontrol agent and the target pathogen by deriving the biocontrol agent from the pathogen. For example, disruption of the *hrp* genes of *Pseudomonas solanacearum* does not affect the ability of the mutant to penetrate host roots and multiply within host tissues, but disruption of the *hrp* genes does affect the ability of the pathogen to cause disease.[1,2] Thus, these avirulent derivatives are adapted to the same ecological niche as the parental strain and can be used to preempt or displace the pathogen from the susceptible host.

II.   BIOLOGICAL CONTROL OF CROWN GALL

Crown gall, a tumorous disease of many dicotyledonous plants, is caused by the soil bacterium, *Agrobacterium tumefaciens*.[3,4] The pathogen penetrates plants at wound sites or root lenticels. During the infection, a segment of DNA (T-DNA) is excised from a large virulence plasmid (Ti plasmid) present in the bacterium and transferred to the host plant where it becomes integrated into the plant cell genome. The integrated DNA encodes enzymes which catalyze the overproduction of auxins and cytokinins in the transformed cells, resulting in unregulated cell division and massive gall formation. In addition, the integrated T-DNA encodes enzymes for the synthesis of opines, low molecular weight molecules which serve as nutrient sources for the pathogen.

In many plants, crown gall can be controlled with *Agrobacterium radiobacter* strain K84, a nonpathogenic bacterium that is closely related to *A. tumefaciens*.[5,6] Biocontrol of crown gall by strain K84 is mediated by the biosynthesis and secretion of a bacteriocin, agrocin 84, in conjunction with the efficient colonization of the host root by the biocontrol agent.[5,7,8] Agrocin 84 biosynthetic genes,[7–9] genes conferring immunity to agrocin 84[10,11] and conjugal transfer capacity genes (*tra*)[11] are located on a 47.7 kb plasmid pAgK84.

Agrocin 84, a di-substituted fraudulent analog of adenosine nucleoside, specifically inhibits pathogenic strains of *A. tumefaciens* with a nopaline type Ti plasmid.[12] Agrocinopines are phosphorylated opines that are synthesized in plant cells which have been transformed with nopaline type T-DNA. The enzymes for the transport and catabolism of agrocinopines by the *Agrobacterium* are encoded on the Ti plasmid. Agrocin 84 is transported into the pathogen by the agrocinopine permease[12,13] and acts by terminating DNA synthesis in the recipient bacterium.[14]

A. TRA– MUTANTS

Transfer of the plasmid from strain K84 to pathogenic strains would result in agrocin 84-resistant pathogens.[8,11] Recombinant DNA techniques were used to construct a strain of *A. radiobacter* (Strain K1026) in which the *tra* region of plasmid pAgK84 was deleted.[15] The ability of the Tra– strain to colonize almond roots and control crown gall was indistinguishable from that of wild type strain K84.[16]

B. ANTIBIOSIS/NICHE COMPETITION

The efficacy of a crown gall biological control agent is not solely dependent upon the production of agrocin 84. Additional mechanisms such as niche competition or production of other agrocins play a role in the suppression of the disease.

A mutant of *A. radiobacter* (Agr–) that was cured of plasmid pAgK84 did not prevent crown gall of tomato stems when coinoculated with a pathogenic strain of *A. tumefaciens*, but reduced infection when plants were inoculated with the mutant 24 hours prior to being inoculated with the pathogen.[17]

Likewise, pre-inoculation of plum with strain K84 reduced the incidence of control crown gall on plants which were subsequently inoculated with either agrocin 84-sensitive strains or agrocin 84-resistant strains of *A. tumefaciens*.[18] The effectiveness of strains K84 was greater against the agrocin 84-sensitive pathogen than against the resistant pathogen.

A mutant of strain K84 which did not produce agrocin 84 also restricted the development of crown gall caused by both the agrocin 84-sensitive and resistant *A.*

tumefaciens. A wild-type strain of K84 was more effective than the Agr– strain in controlling crown gall caused by agrocin 84-sensitive strains, but the levels of control against agrocin 84-resistant strains of *A. tumefaciens* by the wild-type strain K84 and the mutant strain were similar.[18]

Finally, Tra– mutants of the plasmid pAgK84 were transformed into two chromosomal backgrounds of *A. radiobacter* (strains K84 and C58).[19] Levels of agrocin 84 expression were similar, but the K84-derivative effectively controlled crown gall, whereas the C58 transformant harboring the same plasmid was much less proficient at colonizing almond roots and therefore less efficient at controlling crown gall. Differences in the effectiveness of crown gall control were presumably due to differences in colonization.

Surprisingly, increased production of agrocin 84 did not result in increased levels of biocontrol. *Agrobacterium radiobacter* strains carrying a high-copy number mutant of pAgK84 overproduced agrocin 84 *in vitro* but were not more effective than the wild type in controlling crown gall.[19]

The biocontrol of crown gall disease by *A. radiobacter* K84 is primarily mediated through the production of agrocin 84. The results indicate that the production of agrocin-84 greatly enhances the effectiveness of the biological control agents, but competition for colonization sites or nutrients also contributes to the efficacy of control.[17-19]

C.   NON-AGROCIN 84 ANTIBIOSIS

In addition to agrocin 84, *A. radiobacter* strains K84 and K1026 also secrete agrocin 434, a di-substituted cytidine nucleoside which inhibits a broader range of *Agrobacterium* than agrocin 84.[20] Production of agrocin 434 by strain K84 and its derivatives may be an additional factor contributing to the biocontrol of crown gall.

The range of *A. tumefaciens* controlled by *A. radiobacter* could be expanded by further increasing the number of agrocins secreted by strain K84 or by creating a biocontrol strain which produces more effective agrocins.[6,21,22]

III.   BIOLOGICAL CONTROL OF BACTERIAL BLIGHTS

A.   *PSEUDOMONAS SYRINGAE*

1.   *hrp* Mutants

Pseudomonas syringae are opportunistic pathogens of a wide variety of plant species.[23] Individual strains or pathovars (pv.) cause a foliar disease in a limited subset of host plants. During pathogenesis, the bacteria invade the plant tissue through natural openings or wounds. In a susceptible host, the bacteria multiply and disperse throughout the plant tissue and cause water-soaked lesions on leaves and other plant parts. In a resistant reaction on nonhost plants or resistant host plants, the bacteria rapidly elicit the hypersensitive response (HR).[24] The HR of higher plants is characterized by the rapid, localized death of plant cells at the site of pathogen invasion, inhibiting the multiplication and spread of the pathogen within the host tissue.

Lindgren et al.[25] characterized eight insertion mutants of *P. s.* pv. *phaseolicola*, the causative agent of halo blight of bean, which had lost the ability to elicit the HR on nonhost plants such as tobacco and the ability to colonize or cause disease on host plants. The mutations did not alter the growth of the mutants on minimal media. A single recombinant plasmid from a wild-type genomic library was isolated that restored the wild-type phenotype to seven of the insertion mutants, suggesting that the genes affected by the insertions were clustered. The cluster of genes was designated as *hrp* (hypersensitive reaction and pathogenicity).[25-27] *hrp* genes or *hrp*-like mutations have been described in *P. s.* pv. *phaseolicola*,[25] *P. s.* pv. *syringae*,[28,29] *P. s.* pv. *tomato*,[30] *P. s.* pv. *morsprunorum*,[31] *P. s.* pv. *pisi*,[32] *P. s.* pv. *tabaci*,[33] and *P. s.* pv. *glycinea*[33] as well as in several other phytopathogenic bacteria including *Erwinia stewartii*,[34] *Erwinia chrysanthemi*,[35] *Erwinia amylovora*,[36] *Pseudomonas solanacearum*,[37,38] and some pathovars of *Xanthomonas campestris*.[38,39]

Despite the effectiveness of biological control using Hrp– mutants of *P. solanacearum*[40] and *E. chrysanthemi*,[41] the use of avirulent Hrp– mutants of *P. syringae* or *X. campestris* as biocontrol agents has not been reported. Avirulent mutants of *P. syringae* and *X. campestris* have also been characterized with inactivated avirulence genes (*avr*).[42]

2.   Niche Competition

Bacterial speck of tomato has been controlled using copper compounds and/or antibiotics. The effectiveness of the treatments is dependent upon environmental conditions and the development of resistance in the pathogen population. A nonpathogenic Tn5- insertion mutant of *P. s.* pv. *tomato* retained its ability to survive on tomato leaves and reduced the incidence of bacterial speck in the greenhouse.[43] The nonpathogen was further modified by selection for copper resistance. Significantly greater control of bacterial speck was obtained when tomato plants were treated with a combination of a copper bactericide and the copper-resistant *P. s.* pv. *tomato* than when plants were treated with either the nonpathogen or the copper spray alone.

3.   Other

Pseudomonas syringae pv. *savastanoi*, a pathogen of olive and oleander, synthesizes plant hormones causing gall formation on the stems of susceptible plants.[44] The growth of phytohormone-deficient mutants in oleander leaves was unaltered relative to the growth of parental *P. s.* pv. *savastanoi*, but pathogenicity was reduced.[45] Mutants which produced only cytokinin or IAA were less virulent than the wild type. Mutants which did not produce IAA or cytokinin incited necrosis at the inoculation site but did not induce any tissue swelling.

B.   Fire Blight

Fire blight, a necrotic disease of rosaceous plants caused by *Erwinia amylovora*, is especially destructive to apple and pear trees.[46] The disease is currently managed using sprays of streptomycin during the bloom period. Streptomycin-resistant *E. amylovora* have been isolated from orchards in California, Washington, Oregon, and Michigan.[47]

1.   Antibiosis

Erwinia herbicola, a nonpathogenic bacterium which occurs as a common epiphyte on apple and pear trees, produces an antibiotic which is inhibitory to *E. amylovora* and can reduce fire blight incidence when sprayed onto apple blossoms before inoculation with *E. amylovora*. The role of the inhibitory substance was demonstrated by comparing the ability of mutants of *E. herbicola* lacking antibiotic production (Ant–) to control fireblight with the ability of wild-type strains to control fire blight.[48] The Ant– were not as effective as the wild type in controlling plant disease, suggesting that antibiotic production is one of the mechanisms involved in the biological control of fireblight. Again, it was apparent that antibiotic production is not the only mechanism by which *E. herbicola* suppresses fire blight, as the Ant– mutants retained some ability to reduce fire blight incidence on ornamental trees,[48] and disease symptoms caused by an antibiotic-resistant strain of *E. amylovora* were reduced when the pear plants were pre-inoculated with the antibiotic-producing *E. herbicola*.[47]

One potential drawback of *E. herbicola* as a biological control agent of fire blight is that the bacterium does not multiply outside of the area of the fruit where it has been introduced.[49] Therefore, the growth of *E. amylovora* is only restricted in wounds where both bacteria are present. Use of an avirulent antibiotic-producing strain of *E. amylovora* might overcome the ecological limitations of *E. herbicola*.

2.   *hrp* Mutants

Three classes of mutants of *E. amylovora* affected in pathogenicity have been identified using transposon mutagenesis.[50] The first class of mutants were nonpathogenic on all plant species tested and unable to induce the HR on tobacco (*hrp* mutant). Class two mutants were nonpathogenic on apple calli but retained the ability to induce a hypersensitive response on tobacco (*dsp* mutant[disease specificity]). Class three mutants were impaired in EPS (extracellular polysaccharide) production.

IV.   BIOLOGICAL CONTROL OF BACTERIAL SOFT ROT

A number of Gram-negative phytobacterial species belonging to the genera *Erwinia* and *Pseudomonas* cause soft rot diseases on numerous harvested crops and growing plants.[51] These plant pathogenic bacteria produce and secrete a battery of cell wall-degrading enzymes that macerate plant tissues and produce soft rot symptoms.[51] Fluorescent pseudomonads have been evaluated for biological control of soft rot;[52] however, pseudomonads may not be ideally suited as biological control agents under all conditions favoring tuber soft rot because environmental conditions maximizing population of *Pseudomonas* spp. are not identical to those conducive for disease development. In certain environments, an avirulent strain of *Erwinia* which would proliferate under the same conditions as the target pathogen might be a superior biocontrol agent.[53,54]

A. PECTINOLYTIC ENZYMES

Numerous studies with mutants of *E. chrysanthemi*,[55–61] *E. carotovora*[54,62] and *Pseudomonas viridiflava*[63] have demonstrated that pectinolytic enzyme production and export are tightly linked with the ability of these bacteria to elicit soft rot disease. *Erwinia carotovora* subsp. *carotovora* mutants defective in the production of pectate lyase have been used in the biological control of soft rot.[64] However, derivation of avirulent strains through sequential disruption of structural genes encoding pectolytic enzymes has proven to be extremely laborious because of the complexity of pectate lyase expression.

All pathogenic strains of *E. chrysanthemi* and *E. carotovora* produce complex sets of pectate-inducible pectate lyase (pel) isozymes.[51,65] Genetic manipulations of pel production in *E. chrysanthemi* and pel + *Escherichia coli* strains have confirmed that a single Pel isozyme is sufficient to cause extensive maceration in potato tubers.[58,66]

Mutants of *E. chrysanthemi* with directed deletions in all of their known pel genes (ΔPel) were still able to macerate plant tissues, although at reduced rates.[60,67] Additional mutations in either the exopolygalacturonate lyase gene (*pehX*)[68] or the exogalacturonosidase gene (*pelX*)[69] did not diminish the maceration capacity of the pectate lyase-deficient mutant of *E. chrysanthemi*. Kelemu and Collmer[61] constructed an *E. chrysanthemi* mutant with deletions or insertions in all of its known genes coding for pectic enzymes. Although the mutant did not cause pitting on pectate semisolid medium used to detect pectolytic activity in bacteria, the mutant still retained some ability to macerate chrysanthemum tissues. Significant pectate lyase activity was detected in rotting chrysanthemum tissue and in minimal media containing chrysanthemum extracts as the sole carbon source, suggesting that *E. chrysanthemi* produces a set of plant-inducible pectate lyase isozymes as well as a set of pectate-inducible pectate lyase isozymes. However, sole expression of the plant-inducible pectic enzymes is not sufficient to cause systemic disease in whole plants.[60]

One possible alternative for creating avirulent mutants of *Erwinia* would be to disrupt the secretion of pectinolytic enzymes into the plant tissue. *Erwinia* strains that were impaired in the secretion of pectolytic enzymes across the outer membrane (Out−) are avirulent or have greatly reduced virulence.[55,56,59,62,70,71] Out− mutants synthesized normal levels of pectate lyases, polygalacturonase, and cellulases. But most of the cell wall-degrading enzyme activity accumulated within the bacterial periplasmic space rather than in the extracellular medium. Out– mutants of *E. chrysanthemi* were found to be avirulent on the test plant, *Saintpaulia ionantha*[55] and Out– mutants of *E. c.* subsp. *carotovora* were impaired in their ability to macerate potato tuber tissue.[62] The avirulence of Out– mutants of *E. chrysanthemi* suggests that both the pectate-inducible and the plant-inducible pectate lyase isozymes are secreted via the Out pathway.[41,60]

Erwinia carotovora subsp. *betavasculorum* produces and secretes pectolytic enzymes, causing vascular necrosis and root rot of sugar beet.[72] The pathogen can also colonize and macerate potato tubers. *Erwinia carotovora* subsp. *betavasculorum* strain Ecb168 produces an antimicrobial substance(s) that suppresses the growth of the related bacterium *E. c.* subsp. *carotovora* in culture and in wounds of potato tubers. Therefore, an avirulent mutant of *E. c.* subsp. *betavasculorum* may be an effective agent for the biological control of tuber soft rot caused by *E. c.* subsp. *carotovora*.[54]

Out– mutants of *E. c.* subsp. *betavasculorum* had a greatly reduced capacity to macerate potato tubers, but wild-type ability to inhibit the growth of *E. c.* subsp. *carotovora*.[54] In addition, the Out– derivatives of *E. c.* subsp. *betavasculorum* colonized wounds of potato tubers indicating that the Out– phenotype was not associated with a complete loss of ecological fitness. And most important, the Out– mutants suppressed populations of *E. c.* subsp. *carotovora* in wounds of potato and suppressed tuber soft rot.

B.   *hrp* GENES

The effectiveness of *Erwinia* biocontrol agents might be further enhanced by additional manipulations such as modification of *hrp* genes.   The ability of the soft rot pathogens *E. chrysanthemi* and *E. carotovora* to elicit the HR on nonhost plants has been uncertain due to the destructive activity of extracellular enzymes on both host and nonhost tissues. A PelABCE- Out– mutant of *E. chrysanthemi* caused rapid necrosis in tobacco leaves that was typical of that elicited by narrow-host-range pathogens.[41] The *hrp* genes of *E. chrysanthemi* were identified using an *hrp* gene cluster from *E. amylovora*,[35] and mutated using transposon mutagenesis.[41] The *hrp* constructs were transformed into the Out– mutant, and two transformants were identified which did not elicit the HR in tobacco leaves. The mutated *hrp*-containing DNA corresponding to the two mutants was then introduced into a virulent wild-type strain of *E. chrysanthemi*. Introduction of the *hrp* mutation resulted in decreased virulence of both the wild-type and the Out– strains on witloof chicory leaves. *Erwinia chrysanthemi* therefore can produce two independent types of necrosis in plant tissues: a macerative necrosis dependent on the secretion of pectic enzymes and a hypersensitive necrosis dependent on the Hrp system.

C.   SYSTEMIC RESISTANCE

Bacterial cell surface components such as lipopolysaccharide (LPS) or proteins might account for plant-bacterium recognition processes. Two mutants of *E. chrysanthemi* which were resistant to φEC2, an *Erwinia* transducing bacteriophage,[73] had modifications of the LPS structure of the outer membrane and were avirulent on *Saintpaulia ionantha.*[74] One of the avirulent mutants induced systemic resistance in inoculated plants against infection with pathogenic wild-type strains. This study also revealed that the LPS of a previously isolated bacteriocin-resistant mutant was altered and that the avirulent mutant was also capable of inducing systemic resistance to soft rot in inoculated *Saintpaulia* plants.[75]

D. ANTIBIOSIS

Pseudomonas putida strain PP22 inhibits the *in vitro* growth of a range of soft rot bacteria including *Erwinia, Pseudomonas,* and *Xanthomonas.*[76] In *in vivo* studies, strain PP22 suppressed the growth of *E. c.* subsp. *carotovora* on potato tubers and survived on the potato tubers for more than 5 weeks. Studies with Tn-5 mutants have demonstrated that the inhibition of *Erwinia* is distinct from the inhibition of *Pseudomonas*, but the genes responsible for the inhibition were not isolated.

E. *PSEUDOMONAS VIRIDIFLAVA*

In contrast to the complexity of pectate lyase expression in *Erwinia, Pseudomonas fluorescens* strain W5182 and *P. viridiflava*[63] only express a single pectate lyase. Both Pel– mutants and Pel regulatory mutants of *P. viridiflava* fail to macerate plant tissue and fail to induce soft rot in susceptible plants.[63,77]

V. BIOLOGICAL CONTROL OF BACTERIAL WILT

Pseudomonas solanacearum is a soilborne plant pathogenic bacterium that causes lethal wilting diseases in many cultivated and wild plants throughout the world.[78] Over 200 plant species are known hosts of the bacterium and new host–pathogen combinations are continually being described. To date, the means available for controlling the disease are limited. Cultivars of several commercial crops have been developed which are resistant to strains of *P. solanacearum* within a geographical area, but the cultivars may or may not be resistant to strains from other areas. Other disease control approaches are being investigated, due to the extensive variability of the pathogen and the limited availability of resistant cultivars.

Numerous attempts have been made to utilize bacteria which are antagonistic towards *P. solanacearum* or avirulent strains of *P. solanacearum* to control bacterial wilt.[1,2,79,80] Although promising results have been obtained under controlled conditions, none of the biocontrol agents have proven reliable under field conditions. In this respect, Tn5-induced avirulent mutants derived from pathogenic strains are being developed as potential biocontrol agents.

A. *hrp* GENES

Boucher[29] isolated 13 prototrophic mutants using transposon mutagenesis which did not produce any disease symptoms on young susceptible tomato plants. Of the mutants, 9 were affected in their ability to induce a hypersensitive response on a nonhost tobacco plant and therefore carried a mutation in an *hrp* gene.[38,81] A 25 kb DNA plasmid isolated from a wild-type *P. solanacearum* was shown to complement 8 of the 9 *hrp* mutations.[38] Localized mutagenesis of this cosmid led to the identification of the *hrp* gene cluster, as well as a second cluster of pathogenicity genes (*dsp* = disease specificity) located on the right end of the *hrp* cluster, which modulated the aggressiveness of the pathogen on tomato.

The invasiveness of 11 Tn-5-induced avirulent mutants of *P. solanacearum* was compared to the invasiveness of their wild-type parent.[82] Invasiveness was estimated by the ability of bacterial strains to penetrate into and multiply within the root system of a susceptible host. Of the 11 avirulent mutants, 10 were invasive but showed reduced colonization and multiplication relative to the virulent wild type. When susceptible plants were inoculated with mixtures of the avirulent and virulent strains of *P. solanacearum*, the virulent strains were no longer able to efficiently colonize the host tissue.[1] Four of the mutants capable of excluding the virulent wild-type have insertions mapping to *hrp* genes and one had an insertion mapping to a *dsp* gene.

The limited invasiveness of *hrp* mutants of *P. solanacearum* and their ability to exclude pathogenic strains from a susceptible host suggest that they could be used as endophytic biocontrol agents.[40] Three pathogenic strains of *P. solanacearum* were rendered nonpathogenic by insertion of the nontransposable Ω-Km interposon[83] within an *hrp* gene.[40] The ability of the Hrp– mutants to colonize the host tissues was reduced compared with that of a pathogenic strain, but they persisted at low levels in the plant. Upon challenge with pathogenic *P. solanacearum*, there was a reduced incidence of disease in plants pre-inoculated with Hrp-strains when compared with the controls.

Erwinia stewartii causes a vascular wilt of sweet corn and a leaf blight of field corn.[84] Water soaking (*wts*) genes are necessary for pathogenicity and the production of water-soaked lesions.[34,85,86] The *wts* genes were required for lesion formation and wilting in corn but were not required for growth within host plant tissue.[34,86–88] The *wts* genes hybridized with *hrp* genes from *Erwinia* and *Pseudomonas*.[34,85] However, *E. stewartii* differs from other pathogens known to contain *hrp* genes in that wild-type strains do not elicit a hypersensitive response in tobacco.

B. Out– Mutants

Pseudomonas solanacearum produces a variety of virulence factors in addition to the EPS (extracellular polysaccharide) which increase the severity of wilting in susceptible hosts.[89] Mutants defective in the export of major extracellular proteins and enzymes (*eep*) failed to infect susceptible hosts via the roots and disseminate throughout plant tissues.[90]

C. Other

Pseudomonas solanacearum spontaneously loses pathogenicity under certain conditions. Spontaneous loss of pathogenicity is most frequently the result of disruption of the *phc*A gene,[91] but has also been linked to the increased expression of a *trans*-acting factor, *esp*R109 or the presence of a small plasmid, pJTPS1.[92] Homology between *esp*R and pJTPS1 has not been detected.[91]

VI. BIOLOGICAL CONTROL OF FROST INJURY

Frost injury of sensitive plants may be incited by ice nucleation active (Ice+) strains of *P. syringae, P. viridiflava, P. fluorescens, X. campestris* pv. *translucens,* and *E. herbicola,* which commonly inhabit the leaf surfaces of healthy plants.[93] Most frost-sensitive plants have the capacity to supercool to –5 or –6°C before water within the tissues freezes. However, Ice+ bacteria catalyze ice formation at temperatures as warm as –2°C, and leaf surface populations of these bacteria initiate damaging ice formation at temperatures of –1 to –5°C. Frost injury is related to the number of bacterial ice nuclei on the plant at the time of freezing. Treatments that reduce the numbers of Ice+ bacteria on plants reduce the severity of frost injury.

Application of antagonistic, non-ice nucleating strains of *P. syringae* or *E. herbicola* onto uncolonized plant tissue reduced subsequent colonization by Ice+ strains of *P. syringae* and reduced the severity of frost injury.[94,95] Ant– mutants of antibiotic-producing non-ice nucleating strains of *P. syringae* and their parental strain do not differ in their ability to inhibit epiphytic populations of *Ice+ P. syringae,* suggesting that competition for environmental resources was the primary mechanism of antagonism between non-ice nucleating and Ice+ strains of *P. syringae* strains.[96] Therefore, it was hypothesized that an isogenic Ice– mutant of *P. syringae* would occupy the same ecological niche as a parental Ice+ strain and would consequently exhibit more effective competitive exclusion against the parental strain than a genetically dissimilar strain.[53,95] Ice- *P. syringae* mutants, containing internal deletions in the ice nucleation gene, were phenotypically identical to the Ice+ parental strains, except for the inability to nucleate ice. The Ice– mutants of *P. syringae* effectively reduced the populations of isogenic Ice+ *P. syringae* strains on potatoes in the laboratory and in the field.[97] However, the near-isogenic Ice- mutants were not always superior to non-isogenic Ice– *P. syringae* strains in the inhibition of a given Ice+ strain. Effective biological control of a diverse range of Ice+ *P. syringae* might be achieved by using either a mixture of genetically diverse Ice– strains of *P. syringae* or a biological control agent such as *P. fluorescens* strain A506, which occupies a large nutritional niche capable of encompassing the ecological diversity of Ice+ *P. syringae* strains.[95]

VII. CONCLUSIONS

Molecular biology has contributed to our understanding of the mechanism of bacterial pathogenesis and the role of various gene products in plant disease. This has resulted in progress in the development of a new generation of biological control agents whose traits have been modified in a concerted effort to improve the wild type. Accomplishments include: (1) the creation of biological control strains of *A. radiobacter* which control crown gall but are unable to transfer genes encoding bacteriocin resistance to target pathogens;[15] (2) the disruption of pathogenicity genes in *Erwinia*[50] and *Pseudomonas*,[1,54] resulting in biocontrol strains capable of growing within plant tissue and excluding the parental pathogen but incapable of eliciting plant disease in noncompromised hosts; and (3) derivation of Ice– strains of epiphytic bacteria which may reduce the severity of frost injury caused by Ice+ strains.[94,97]

As expected, the first generation of engineered biocontrol organisms have modifications in single pathogenicity factors or within a single gene. To paraphrase Coleridge, "Beware of a man with a single idea." For example, *Bacillus thuringiensis* toxin has been proven effective in numerous insect control systems, but resistance is rapidly becoming a problem.[98] Plant breeders have overcome the rapid breakdown of disease resistance by pyramiding major and minor resistance genes within a single plant or within a single field.[99] Resistance may still appear in the pathogen population, but it develops more slowly. The same analogy could be applied to biocontrol strategies. An agent with a full artillery would pose more of a threat to a pathogen than an agent with a single weapon. Although a strain of *A. radiobacter* which does not produce agrocin 84 provides some protection against crown gall, a bacteriocin-producing strain provides significantly more.[17] Alternatively, a battery of biocontrol agents could be released to combat a single pathogen. In either situation, the probability of success of biocontrol will be greater, as there would be less chance that the pathogen could overcome a diversity of biological control agents.

REFERENCES

1. Trigalet, A. and Trigalet-Demery, D., Use of avirulent mutants of *Pseudomonas solanacearum* for the biological control of bacterial wilt of tomato plants, *Physiol. Mol. Plant Pathol.*, 36, 27, 1990.
2. Trigalet, A., Frey, P., and Trigalet-Demery, D., Biological control of bacterial wilt caused by *Pseudomonas solanacearum*: state of the art and understanding, *Bacterial Wilt: The Disease and its Causative Agent, Pseudomonas solanacearum*, Hayward, A. C. and Hartman, G. L., Eds., CAB International, Wallingford, CT, 1994, 225.
3. Zambryski, P., Tempe, J., and Schell, J., Transfer and function of T-DNA genes from *Agrobacterium* Ti and Ri plasmids in plants, *Cell*, 56, 193, 1989.
4. Winans, S., Two-way chemical signaling in *Agrobacterium*-plant interactions, *Microbiol. Rev.*, 56, 12, 1992.
5. Kerr, A. and Htay, K., Biological control of crown gall through bacteriocin production, *Physiol. Plant Pathol.*, 4, 37, 1974.
6. Thomson, J. A., The use of agrocin-producing bacteria in the biological control of crown gall, *Innovative Approaches to Plant Disease Control*, Chet, I., Ed., John Wiley & Sons, New York, 1987, 213.
7. Ellis, J. G., Kerr, A., van Montagu, M., and Schell, J., *Agrobacterium*: genetic studies on agrocin 84 production and the biological control of crown gall, *Physiol. Plant Pathol.*, 15, 311, 1979.

8. Kerr, A., Biological control of crown gall through production of agrocin 84, *Plant Dis.*, 64, 25, 1980.

9. Wang, C. L., Farrand, S. K., and Hwang, I., Organization and expression of the genes on pAgK84 that encode production of agrocin 84, *Mol. Plant-Microbe Interact.*, 7, 472, 1994.

10. Ryder, M. H., Slota, J., Scarim, A., and Farrand, S. K., Genetic analysis of agrocin 84 production and immunity in *Agrobacterium* spp., *J. Bacteriol.*, 169, 4184, 1987.

11. Farrand, S. K., Slota, J. E., Shim, J. S., and Kerr, A., Tn5 insertions in the agrocin 84 plasmid: the conjugal nature of pAGk84 and the locations of determinants for transfer- and agrocin 84 production, *Plasmid*, 13, 106, 1985.

12. Ellis, J. G. and Murphy, P. J., Four new opines from crown gall tumors — their detection and properties, *Mol. Gen. Genet.*, 181, 36, 1981.

13. Hayman, G. T. and Farrand, S. K., Characterization and mapping of the agrocinopine-agrocin 84 locus on the nopaline Ti plasmid pTiC58, *J. Bacteriol.*, 170, 1759, 1988.

14. Murphy, P. J. and Roberts, W. P., A basis for agrocin 84 sensitivity in *Agrobacterium radiobacter*, *J. Gen. Microbiol.*, 114, 207, 1979.

15. Jones, D. A., Ryder, M. H., Clare, B. G., Farrand, S. K., and Kerr, A., Construction of a Tra– deletion mutant of pAgK84 to safeguard the biological control of crown gall, *Mol. Gen. Genet.*, 212, 207, 1988.

16. Jones, D. A. and Kerr, A., *Agrobacterium radiobacter* strain K1026, a genetically engineered derivative of strain K84, for biological control of crown gall, *Plant Dis.*, 73, 15, 1989.

17. Cooksey, D. A. and Moore, L. W., Biological control of crown gall with an agrocin mutant of *Agrobacterium radiobacter*, *Phytopathology*, 72, 919, 1982.

18. López, M. M., Gorris, M. T., Salcedo, C. I., Montojo, A. M., and Miró, M., Evidence of biological control of *Agrobacterium tumefaciens* strains sensitive and resistant to agrocin 84 by different *Agrobacterium radiobacter* strains on stone fruit trees, *Appl. Environ. Microbiol.*, 55, 741, 1989.

19. Shim, J. S., Farrand, S. K., and Kerr, A., Biological control of crown gall: construction and testing of new biocontrol agents, *Phytopathology*, 77, 463, 1987.

20. Donner, S. C., Jones, D. A., McClure, N. C., Rosewarne, G. M., Tate, M. E., Kerr, A., Fajardo, N. N., and Clare, B. G., Agrocin 434, a new plasmid encoded agrocin from the biocontrol *Agrobacterium* strains K84 and K1026, which inhibits biovar 2 agrobacteria, *Physiol. Mol. Plant Pathol.*, 42, 185, 1993.

21. Hendson, M., Askjaer, L., Thomson, J. A., and van Montagu, M., Broad-host range agrocin of *Agrobacterium tumefaciens*, *Appl. Environ. Microbiol.*, 45, 1526, 1983.

22. Webster, J., Dos Santos, M., and Thomson, J. A., Agrocin-producing A*grobacterium tumefaciens* strain active against grapevine isolates, *Appl. Environ. Microbiol.*, 52, 217, 1986.

23. Schroth, M. N., Hildebrand, D. C., and Starr, M. P., Phytopathogenic members of the genus *Pseudomonas*, in *The Procaryotes*, Starr, M. P., Stolp, H., Trüper, H. G., Ballows, A., and Schlegal, H. G., Eds., Springer-Verlag, Berlin, 1981, 701.

24. Klement, Z., Hypersensitivity, in *Phytopathogenic Prokaryotes*, Vol. 2, Mount, M. S. and Lacy, G. H., Eds., Academic Press, New York, 1982, 149.

25. Lindgren, P. B., Peet, R. C., and Panopoulos, N. J., Gene cluster of *Pseudomonas syringae* pv. *"phasiolicola"* controls pathogenicity of bean plants and hypersensitivity on nonhost plants, *J. Bacteriol.*, 168, 512, 1986.

26. Lindgren, P. B., Fredrick, R., Govindarajan, A. G., Panopoulos, N. J., Staskawicz, B. J., and Lindow, S. E., an ice nucleation reporter gene system: identification of inducible pathogenicity genes in *Pseudomonas syringae* pv. *phasiolicola*, *EMBO J.*, 8, 1291, 1989.

27. Rahme, L. G., Mindrinos, M. N., and Panopoulos, N. J., The genetic and transcriptional organization of the *hrp* cluster of *Pseudomonas syringae* pathovar *phasiolicola*, *J. Bacteriol.*, 170, 5479, 1991.

28. Huang, H.-C., Schuurink, R., Denny, T. P., Atkinson, M. M., Baker, C. J., Yucel, I., Hutcheson, S. W., and Collmer, A., Molecular cloning of a *Pseudomonas syringae* pv. *syringae* gene cluster that enables *Pseudomonas fluorescens* to elicit the hypersensitive response in tobacco plants, *J. Bacteriol.*, 170, 4748, 1988.

29. Mukhopadhyay, P., Williams, J., and Mills, D., Molecular analysis of a pathogenicity locus in *Pseudomonas syringae* pv. *syringae*, *J. Bacteriol.*, 170, 5479, 1988.

30. Cuppels, D. A., Generation and characterization of Tn5 insertion mutations in *Pseudomonas syringae* pv. *tomato*, *Appl. Environ. Microbiol.*, 51, 323, 1986.

31. Liang, L. S. and Jones, A. L., Organization of the *hrp* gene cluster and nucleotide sequence of the *hrpL* gene from *Pseudomonas syringae* pv. *morsprunorum*, *Phytopathology*, 85, 118, 1995.

32. Malik, A. N., Vivian, A., and Taylor, J. D., Isolation and partial characterization of three classes of mutants in *Pseudomonas syringae* pathovar *pisi* with altered behavior towards their host, *Pisum sativum*, *J. Gen. Microbiol.*, 133, 2393, 1987.

33. Lindgren, P. B., Panopoulos, N. J., Staskawicz, B. J., and Dahlbeck, D., Genes required for pathogenicity and hypersensitivity are conserved and interchangeable among pathovars of *Pseudomonas syringae*, *Mol. Gen. Genet.*, 211, 499, 1988.

34. Coplin, D. L., Frederick, R. D., Majerczak, D. R., and Tuttle, L. D., Characterization of a gene cluster that specifies pathogenicity in *Erwinia stewartii*, *Mol. Plant-Microbe Interact.*, 5, 81, 1992.

35. Laby, R. J. and Beer, S. V., Hybridization and functional complementation of the *hrp* gene cluster from *Erwinia amylovora* strain Ea321 with DNA of other bacteria, *Mol. Plant-Microbe Interact.*, 5, 412, 1992.

36. Bauer, D. W. and Beer, S. V., Cloning of a gene from *Erwinia amylovora* involved in induction of hypersensitivity and pathogenicity, *Plant Pathogenic Bacteria*, Civerolo, E. L., Collmer, A., Davis, R. E., and Gillaspie, A. G., Eds., Martinus Nijhoff, Boston, 1987, 425.

37. Boucher, C. A., Barberis, P. A., Trigalet, A. P., and Demery, D. A., Transposon mutagenesis of *Pseudomonas solanacearum*: isolation of Tn5-induced mutants, *J. Gen. Microbiol.*, 131, 2449, 1985.

38. Boucher, C. A., Van Gijsegem, F., Barberis, P. A., Arlat, M., and Zischek, C., *Pseudomonas solanacearum* genes controlling both pathogenicity on tomato and hypersensitivity on tobacco are clustered, *J. Bacteriol.*, 169, 5626, 1987.

39. Bonas, U., Schulte, R., Fenselau, S., Minsavage, G. V., Staskawicz, B. J., and Stall, R. E., Isolation of a gene cluster from *Xanthomonas campestris* pv. *vesicatoria* that determines pathogenicity and the hypersensitive response on pepper and tomato, *Mol. Plant-Microbe Interact.*, 4, 81, 1991.

40. Frey, P., Prior, P., Marie, C., Kotoujansky, A., Trigalet-Demery, D., and Trigalet, A., Hrp– mutants of *Pseudomonas solanacearum* as potential biocontrol agents of tomato bacterial wilt, *Appl. Environ. Microbiol.*, 60, 3175, 1994.

41. Bauer, D. W., Bogdanove, A. J., Beer, S. V., and Collmer, A., *Erwinia chrysanthemi hrp* genes and their involvement in soft rot pathogenesis and elicitation of the hypersensitive response, *Mol. Plant-Microbe Interact.*, 7, 573, 1994.

42. Lorang, J. M., Shen, H., Kobayashi, D., Cooksey, D., and Keen, N. T., *avrA* and *avrE* in *Pseudomonas syringae* pv. *tomato* PT23 play a role in virulence on tomato plants, *Mol. Plant-Microbe Interact.*, 7, 508, 1994.

43. Cooksey, D. A., Reduction of infection by *Pseudomonas syringae* pv. *tomato* using a nonpathogenic, copper-resistant strain combined with a copper bactericide, *Phytopathology*, 78, 601, 1988.

44. Surico, G., Comai, L., and Kosuge, T., Pathogenicity of strains of *Pseudomonas syringae* pv. *savastanoi* and their indoleacetic acid-deficient mutants on olive and oleander, *Phytopathology*, 74, 490, 1984.

45. Iacobellis, N., Sisto, A., Surico, G., Evidente, A., and DiMaio, E., Pathogenicity of *Pseudomonas syringae* subsp. *savastanoi* mutants defective in phytohormone production, *J. Phytopathology*, 140, 238, 1994.

46. Schroth, M. N., Thomson, S. V., and Hildebrand, D. C., Epidemiology and control of fireblight, *Annu. Rev. Phytopathol.*, 12, 389, 1974.

47. Wodzinski, R. S., Umhostz, T. E., Rundle, J. R., and Beer, S. V., Mechanisms of inhibition of *Erwinia amylovora* by *Erw. herbicola in vitro* and *in vivo*, *J. Appl. Bact.*, 76, 22, 1994.

48. Vanneste, J. L., Yu, J., and Beer, S. V., Role of antibiotic production by *Erwinia herbicola* Eh252 in biological control of *Erwinia amylovora*, *J. Bacteriol.*, 174, 2785, 1992.

49. Beer, S. V., Rundle, J. R., and Norelli, J. L., Recent progress in the development of biological control for fire blight — a review, *Acta Hortic.*, 151, 195, 1984.

50. Vanneste, J. L., Paulin, J. P., and Expert, D., Bacteriophage Mu as a genetic tool to study *Erwinia amylovora* pathogenicity and hypersensitive reaction on tobacco, *J. Bacteriol.*, 172, 932, 1990.

51. Collmer, A. and Keen, N. T., The role of pectic enzymes in plant pathogenesis, *Annu. Rev. Phytopathol.*, 24, 383, 1986.

52. Xu, G.-W. and Gross, D. C., Selection of fluorescent pseudomonads antagonistic to *Erwinia carotovora* and suppressive of potato seed piece decay, *Phytopathology*, 76, 414, 1986.

53. Wilson, M. and Lindow, S., Release of recombinant microorganisms, *Annu. Rev. Microbiol.*, 47, 913, 1993.

54. Costa, J. M. and Loper, J. E., Derivation of mutants of *Erwinia carotovora* subsp. *betavasculorum* deficient in export of pectolytic enzymes with potential for biological control of potato soft rot, *Appl. Environ. Microbiol.*, 60, 2278, 1994.

55. Andro, T., Chambost, J. P., Kotoujansky, A., Cattaneo, J., Bertheau, Y., Barras, F., Van Gijsegem, F., and Coleno, A., Mutant of *Erwinia chrysanthemi* defective in secretion of pectinase and cellulase, *J. Bacteriol.*, 160, 1199, 1984.

56. Thurn, K. K. and Chatterjee, A. K., Single-site chromosomal Tn5 insertions affect the export of pectolytic and cellulolytic enzymes in *Erwinia chrysanthemi* EC16, *Appl. Environ. Microbiol.*, 50, 894, 1985.

57. Hugouvieux-Cotte-Pattat, N., Reverchon, S., Condemine, G., and Robert-Baudouy, J., Regulatory mutations affecting the synthesis of pectate lyase in *Erwinia chrysanthemi*, *J. Gen. Microbiol.*, 132, 2099, 1986.

58. Payne, J. H., Schoedel, C., Keen, N. T., and Collmer, A., Multiplication and virulence in plant tissues of *Escherichia coli* clones producing pectate lyase isozymes PLb and PLe at high levels and of an *Erwinia chrysanthemi* mutant deficient in PLe, *Appl. Environ. Microbiol.*, 53, 2315, 1987.

59. Beaulieu, C. and Van Gijsegem, F., Pathogenic behavior of several mini-mu-induced mutants of *Erwinia chrysanthemi* on different plants, *Mol. Plant-Microbe Interact.*, 5, 340, 1992.

60. Beaulieu, C., Boccara, M., and Van Gijsegem, F., Pathogenic behavior of pectinase-defective *Erwinia chrysanthemi* mutants on different plants, *Mol. Plant-Microbe Interact.*, 6, 197, 1993.

61. Kelemu, S. and Collmer, A., *Erwinia chrysanthemi* EC16 produces a second set of plant-inducible pectate lyase isozymes, *Appl. Environ. Microbiol.*, 59, 1756, 1993.

62. Murata, H., Fons, M., Chatterjee, A., Collmer, A., and Chatterjee, A. K., Characterization of transposon insertion Out– mutants of *Erwinia carotovora* subsp. *carotovora* defective in enzyme export and of a DNA segment that complements *out* mutations in *E. carotovora* subsp. *carotovora*, *E. carotovora* subsp. *atroseptica*, and *Erwinia chrysanthemi*, *J. Bacteriol.*, 172, 2970, 1990.

63. Liao, C.-H., Hung, H.-Y., and Chatterjee, A. K., An extracellular pectate lyase is the pathogenicity factor of the soft-rotting bacterium *Pseudomonas viridiflava*, *Mol. Plant-Microbe Interact.*, 1, 199, 1988.

64. Stromber, V. K., Orvos, D. R., and Scanferlato, V. S., *In planta* competition among cell-degrading enzyme mutants and wild-type strains of *Erwinia carotovora*, *Proc. 7th Int. Conf. Plant Pathogenic Bacteria*, Klement, Z., Ed., Akedimia Kiado, Budapest, 1990, 721.

65. Reverchon, S., Van Gijsegem, F., Rouve, M., Kotoujansky, A., and Robert-Baudouy, J., Organization of a pectate lyase gene family in *Erwinia chrysanthemi*, *Gene*, 49, 215, 1986.

66. Boccara, M., Diolez, A., Rouve, M., and Kotoujansky, A., The role of individual pectate lyases of *Erwinia chrysanthemi* strain 3937 in pathogenicity on *saintpaulia* plants, *Physiol. Mol. Plant Pathol.*, 33, 95, 1988.

67. Ried, J. L. and Collmer, A., Construction and characterization of an *Erwinia chrysanthemi* mutant with directed deletions in all of the pectate lyase structural genes, *Mol. Plant-Microbe Interact.*, 1, 32, 1988.

68. He, S. Y. and Collmer, A., Molecular cloning, nucleotide sequence, and marker exchange mutagenesis of the exo-poly-α-D-galacturonosidase-encoding *pehX* gene of *Erwinia chrysanthemi* EC16, *J. Bacteriol.*, 172, 4988, 1990.

69. Brooks, A. D., He, S. Y., Gold, S., Keen, N. T., Collmer, A., and Hutcheson, S. W., Molecular cloning of the structural gene for exopolygalacturonate lyase from *Erwinia chrysanthemi* EC16 and characterization of the enzyme product, *J. Bacteriol.*, 172, 6950, 1990.

70. Pirhonen, M., Saarilahti, H., Karlsson, M. B., and Palva, E. T., Identification of pathogenicity determinants of *Erwinia carotovora* subsp. *carotovora* by transposon mutagenesis, *Mol. Plant-Microbe Interact.*, 4, 276, 1991.

71. Hinton, J. C. D., Sidebotham, J. M., Hyman, L. J., and Pérombelon, M. C. M., Isolation and characterization of transposon-induced mutants of *Erwinia carotovora* subsp. *atroseptica* exhibiting reduced virulence, *Mol. Gen. Genet.*, 217, 141, 1989.

72. Thomson, S. V., Hildebrand, D. C., and Schroth, M. N., Identification and nutritional differentiation of the *Erwinia* sugar beet pathogen from members of *Erwinia carotovora* and *Erwinia chrysanthemi*, *Phytopathology*, 71, 1037, 1977.

73. Resibois, A., Colet, M., Faelen, M., Schoonejans, E., and Toussaint, A., φEC2, a new generalized transducing phage of *Erwinia chrysanthemi*, *Virology*, 137, 102, 1984.

74. Schoonejans, E., Expert, D., and Toussaint, A., Characterization and virulence properties of *Erwinia chrysanthemi* lipopolysaccharide-defective, φEC2-resistant mutants, *J. Bacteriol.*, 169, 4011, 1987.

75. Expert, D. and Toussaint, A., Bacteriocin-resistant mutants of *Erwinia chrysanthemi*: possible involvement of iron acquisition in phytopathogenicity, *J. Bacteriol.*, 163, 221, 1985.

76. Liao, C. H., Antagonism of *Pseudomonas putida* strain PP22 to phytopathogenic bacteria and its potential use as a biocontrol agent, *Plant Disease*, 73, 223, 1989.

77. Liao, C.-H., McCallus, D. E., and Fett, W. F., Molecular characterization of two gene loci required for production of the key pathogenicity factor pectate lyase in *Pseudomonas viridiflava*, *Mol. Plant-Microbe Interact.*, 7, 391, 1994.

78. Hayward, A. C., Biology and epidemiology of bacterial wilt caused by *Pseudomonas solanacearum*, *Annu. Rev. Phytopathol.*, 29, 65, 1991.

79. El-Abyad, M. S., El-Sayed, M. A., El-Shanshoury, A.-R., and El-Batanouny, N. H., Inhibitory effects of UV mutants of Strept*omyces corchorusii* and *Streptomyces spiroverticillatus* on bean and banana wilt pathogens, *Can. J. Bot.*, 71, 1080, 1993.

80. Chen, W. and Echandi, E., Effects of avirulent bacteriocin-producing strains of *Pseudomonas solanacearum* on the control of bacterial wilt of tobacco, *Plant Pathol.*, 33, 245, 1984.

81. Willis, D. K., Rich, J. J., and Hrabak, E. M., *hrp* genes of phytopathogenic bacteria, *Mol. Plant-Microbe Interact.*, 4, 132, 1991.

82. Trigalet, A. and Demery, D., Invasiveness in tomato plants of Tn5-induced avirulent mutants of *Pseudomonas solanacearum*, *Physiol. Mol. Plant Pathol.*, 28, 423, 1986.

83. Fellay, R., Frey, J., and Krish, H., Interposon mutagenesis of soil and water bacteria: a family of DNA fragments designed for in *vitro* insertional mutagenesis of gram-negative bacteria, *Gene*, 52, 147, 1987.

84. Pepper, E. H., *Stewart's Bacterial Wilt of Corn*. Monograph No. 4, American Phytological Society, St. Paul, 1967.

85. Frederick, R. D., Majerczak, D. R., and Coplin, D. L., *Erwinia stewartii* WtsA, a positive regulator of pathogenicity gene expression, is similar to *Pseudomonas syringae* pv. *phasiolicola* HrpS, *Mol. Microbiol.*, 9, 477, 1993.

86. Coplin, D. L., Frederick, R. D., and Majerczak, D. R., New pathogenicity loci in *Erwinia stewartii* identified by random Tn5 mutagenesis and molecular cloning, *Mol. Plant-Microbe Interact.*, 5, 266, 1992.

87. Coplin, K. L. and Majerczak, D. R., Extracellular polysaccharide genes in *Erwinia stewartii*: directed mutagenesis and complementation analysis, *Mol. Plant-Microbe Interact.*, 3, 286, 1990.

88. Dolph, P. J., Majerczak, D. R., and Coplin, D. L., Characterization of a gene cluster for exopolysaccharide biosynthesis and virulence in *Erwinia stewartii*, *J. Bacteriol.*, 170, 865, 1988.

89. Denny, T. P., Carney, B. F., and Schell, M. A., Inactivation of multiple virulence genes reduces the ability of *Pseudomonas solanacearum* to cause wilt symptoms, *Mol. Plant-Microbe Interact.*, 3, 293, 1990.

90. Kang, Y., Huang, J., Mao, G., He, L., and Schell, M. A., Dramatically reduced virulence of mutants of *Pseudomonas solanacearum* defective in export of extracellular proteins across the outer membrane, *Mol. Plant-Microbe Interact.*, 7, 370, 1994.

91. Brumbley, S. M., Carney, B. F., and Denny, T. P., Phenotype conversion in *Pseudomonas solanacearum* due to spontaneous inactivation of PhcA, a putative LysR transcriptional regulator, *J. Bacteriol.*, 175, 5477, 1993.

92. Negishi, H., Yamada, T., Shiraishi, T., Oku, H., and Tanaka, H., *Pseudomonas solanacearum*: plasmid pJTPS1 mediates a shift from the pathogenic to nonpathogenic phenotype, *Mol. Plant-Microbe Interact.*, 6, 203, 1993.

93. Hu, N.-T., Hung, M.-N., Chiou, S.-J., Tang, F., Chiang, D.-C., Huang, H.-Y., and Wu, C.-Y., Cloning and characterization of a gene required for the secretion of extracellular enzymes across the outer membrane by *Xanthomonas campestris* pv. *campestris*, *J. Bacteriol.*, 174, 2679, 1992.

94. Lindemann, J. and Suslow, T. V., Competition between ice nucleation-active wild type and ice nucleation-deficient deletion mutant strains of *Pseudomonas syringae* and *P. fluorescens* Biovar I and biological control of frost injury on strawberry blossoms, *Phytopathology*, 77, 882, 1987.

95. Wilson, M. and Lindow, S. E., Ecological similarity and coexistence of epiphytic ice-nucleating (Ice+) *Pseudomonas syringae* strains and a non-ice-nucleating (Ice–) biological control agent, *Appl. Environ. Microbiol.*, 60, 3128, 1994.

96. Lindow, S. E., Lack of correlation in *vitro* antibiosis with antagonism of ice nucleation active bacteria on leaf surfaces by non-ice nucleation active bacteria, *Phytopathology*, 78, 444, 1988.

97. Lindow, S. E., Ice– strains of *Pseudomonas syringae* introduced to control ice nucleation active strains on potato, in *Biological Control of Plant Diseases: Progress and Challenges for the Future*, Tjamos, E. S., Papavizas, G. C., and Cook, R. J., Eds., Plenum Press, New York, 1992, 169.

98. Gould, F., Martinez-Ramirez, A., Anderson, A., Ferre, J., Silva, F. J., and Moar, W. F., Broad-spectrum resistance to *Bacillus thuringiensis* toxins in *Heliothis virescens*, *Proc. Natl. Acad. Sci. U.S.A.*, 89, 7986, 1992.

99. Wolfe, M. S. and Gessler, C., The use of resistance genes in breeding epidemiological considerations, in *Genes Involved in Plant Defense*, Boller, T. and Meins, F., Eds., Springer-Verlag, Vienna, 1992, 3.

4 Molecular Biology of Fungal Diseases

Christine D. Smart and Dennis W. Fulbright

TABLE OF CONTENTS

I.   INTRODUCTION

The biological control of fungal diseases of plants is one of the great challenges left for the discipline of plant pathology. Not only will understanding the biology be a worthy challenge, but altering the paradigm of current disease management strategies which rely heavily on chemical control will take time and patience for products to be marketed and delivered to an expectant but somewhat skeptical clientele. For biological control advocates to succeed in breaking down the barriers, a thorough understanding of the biological systems must be attained. The reasons for success must be clearly established and understood in a scientific and rational manner if we are to deal with the inevitable periodic failure of control. For example, our understanding of the biological control of crown gall by *Agrobacterium radiobacter* strain K84 is such that we can predict when the biocontrol might not be effective and in some cases prescribe genetic strategies to preclude some of the observed or expected failures.[1] Such strategies can only be designed when the basic elements of the control system are well understood.

Chemical control of viral and bacterial diseases has been rare, and disease management has mainly been achieved through cultural management and/or host resistance. Fungal diseases, however, have been managed primarily and successfully through chemical application. The rationale behind traditional biocontrol strategies of fungal diseases has been to control those relatively few diseases that were not managed with chemicals. Due to environmental and food safety concerns, the driving force of biocontrol may soon be food industry regulations that call for the elimination of historically successful fungicides, allowing once easily managed diseases to remerge as potential threats. Surely, at that time, biocontrol agents and techniques will be tapped to protect crops from the ensuing onslaught.

II.   HYPOVIRULENCE

In this chapter we will concentrate on "hypovirulence", a biocontrol system studied due to its success in altering to some extent chestnut blight, an unmanageable forest disease. Hypovirulence has been variously described as a form of induced resistance, antibiosis, and hyperparasitism. Through recent studies we now know that hypovirulence is simply a pathogen phenotype where virulence is reduced.[2] When the hypovirulence phenotype is cytoplasmically transferred to virulent strains, hypovirulence has an opportunity to spread horizontally through the pathogen population. It is in this manner that hypovirulence is thought to alter the virulence of a pathogen population.

Hypovirulence has been frequently linked to growth rate and morphology of the pathogen. Almost all of the hypovirulent strains of *Cryphonectria parasitica*,[3-5] *Ophiostoma ulmi*,[6] *Sclerotinia sclerotiorum*,[7] *Gaeumannomyces graminis* var. *tritici*,[8] and *Magnaporthe grisea*[9] show altered growth phenotypes whether induced by cytoplasmic-borne or nuclear-borne genes. Since growth can be modulated by many different factors, it is not surprising to find at least three different causes for hypovirulence in phytopathogenic fungi; mycovirus infection, mitochondrial DNA mutations, and nuclear genome mutations. When hypovirulence is observed in natural biocontrol situations, the "infectious" or transmissible agents responsible for the phenotype have been either viral and/or mitochondrial.

A.   MYCOVIRUSES AND PHYTOPATHOGENIC FUNGI

Although viruses whose hosts are in the plant and animal kingdoms have been studied since the late 19th century, mycoviruses were not described until 1962.[10]

Since that time, a great deal has been learned about this group of viruses. The genome of most mycoviruses is composed of double-stranded RNA (dsRNA), and these viruses have been detected in all classes of fungi.[10] While the majority of virus infections of fungi are symptomless,[11] some mycoviruses are responsible for the production of toxins[12] or "killer factors"[13] that appear to act similarly to prokaryotic bacteriocins. Fungal isolates that produce killer factors inhibit the growth of closely related fungal strains not harboring the mycovirus since the mycovirus is also responsible for killer factor immunity. Unless they are somehow an artifact of the laboratory environment, killer factors must be involved in antibiosis in natural systems, yet little in the way of biocontrol has been subscribed to killer factor systems. Still, this system is worth continued exploration in light of the success of antibiosis in prokaryotic systems.[14] Plant pathologists studying biocontrol strategies should keep abreast of killer factor research frequently published in genetics and virology journals, as it may be exploitable in light of molecular biological techniques.[15]

When the etiology of hypovirulence abnormalities is mycoviral, the symptoms can be subtle or extreme. In one of the best studied pathosystems where mycovirus infection plays a role in the outcome of the disease, the chestnut blight/hypovirulence system, strains may be so little affected that it can be difficult to determine if they are infected[16] or they may be changed so dramatically as to make it difficult to determine the species of the mycovirus-infected isolate.[17] A closer look at two pathosystems where mycovirus infection leads to hypovirulence follows.

1.   D-factors: Hypovirulence in *Ophiostoma ulmi*
       and *O. novo-ulmi*

The ascomycete *Ophiostoma ulmi*, the causal organism of Dutch elm disease, is a devastating disease of elm species in Europe, North America and Southwest Asia. There are two major subgroups of this fungus; one is a highly pathogenic, aggressive strain, which is quickly replacing the other, a nonaggressive strain.[18] It is now thought the the newer aggressive strain is a new species, *O. novo-ulmi*,[19] that is outcompeting the nonaggressive strain. Superimposed over the pathogenicity coded for by the nuclear genome are cytoplasmically transmissible disease factors, called d-factors, which have been identified in both *O. ulmi* and *O. novo-ulmi*. Diseased isolates are characterized by the presence of multiple segments of dsRNA, production of sectors of weak or abnormal growth, reduction in viability of conidia, and reduction in the ability to sexually reproduce.[6]

The specific dsRNA segment(s) responsible for the symptoms associated with diseased isolates has been difficult to determine since unencapsidated dsRNA segments occur in both diseased and healthy *O. ulmi* and *O. novo-ulmi* strains. However, transmission of the diseased state to a healthy strain was correlated with the transmission of 10 dsRNA segments.[18] dsRNA in diseased strains has also been found to copurify with the mitochondria (see Section II. B. 2, below), which are deficient in cytochrome aa_3.[20] It is not known if the dsRNA is directly responsible for this deficiency, but suppression of respiration may be a mechanism by which fungal growth is reduced.[20]

The possible use of dsRNA d-factors as a biocontrol agent is very exciting; however, there are at least two barriers to overcome. First, since the dsRNA is not transmitted through the sexual cycle of the fungus, a large number of uninfected ascospores capable of causing disease will be produced.[21] Also, the large number of vegetative compatibility groups within *O. ulmi* limits the spread of d-factors through hyphal anastomosis. Even with these limitations, the number of naturally occurring diseased strains is quite high (up to 40% in some locations in Europe), leading researchers to believe that biocontrol may be possible.[21]

2.   dsRNA in *Cryphonectria parasitica*

a.   Chestnut Blight

The American chestnut tree (*Castanea dentata*) was once a major component of the hardwood forests in eastern North America. In the early 1900s, over 3 billion chestnut trees were killed after the fungal pathogen *Cryphonectria parasitica*, the causal organism of chestnut blight, was introduced from Asia.[17] The disease was transported to Italy in the 1930s, where *C. parasitica* attacked the European chestnut (*Castanea sativa*).[22] Unlike the situation in North America, some European trees survived, even though they were infected by the chestnut blight fungus. *C. parasitica* strains isolated from these trees were less virulent than other strains and were termed hypovirulent.[23] Surprisingly, lethal cankers could be converted to nonlethal cankers by inoculating them with mycelia from the hypovirulent strains.[24] It was determined that this phenomenon was due to transmission of a cytoplasmic factor from the hypovirulent strain to the virulent strain through hyphal anastomosis. In 1977, Day et al.[25] reported that hypovirulent forms of *C. parasitica* contained dsRNA molecules, while virulent strains did not. A causal relationship between the presence of dsRNA and hypovirulence was implied, and it was suggested that the dsRNA was of viral origin. When researchers noted the similarity between the Michigan and European situations, surviving American chestnut trees in Michigan were tested for hypovirulent isolates, and hypovirulent strains were found in 1976.[26] Surveys indicated that several blighted but recovering stands contained hypovirulent isolates with dsRNA.[5]

dsRNA-containing hypovirulent isolates of *C. parasitica* have been collected from many locations throughout France and Italy,[27] as well as in several locations in North America. The dsRNA molecules in *C. parasitica* have been associated with virus-like particles (VLPs) in the cytoplasm of hypovirulent *C. parasitica* strains.[28-29] Similar vesicles have been found in virulent strains; however they do not contain dsRNA. Fungal polysaccharides have been detected in vesicles from virulent and hypovirulent strains, suggesting that these vesicles could function in fungal cell wall synthesis.[30] An RNA-dependent RNA polymerase associated with dsRNA- containing-vesicles has been identified and characterized and appears to function in the replication of the dsRNA.[30] It is thought that the replication strategy of the dsRNA molecules in *C. parasitica* is similar to that of positive-strand RNA viruses.[31]

Hypovirulence-associated dsRNA molecules from Europe and North America have many similarities, including: cytoplasmic transmissibility, correlation with reduction in virulence, association with membrane-bound particles, and effects on fungal growth and morphology.[2] However, major differences have been identified. The most obvious morphological differences between European and North American hypovrulent strains were based on pigmentation and sporulation differences. The North American hypovirulent strains are fully pigmented, while European strains generally (but not always) lack pigmentation; sporulation was not as noticeably suppressed in North American hypovirulent strains as in European isolates. dsRNA genomes from several European strains cross-hybridized with each other upon northern analysis; however, they did not cross-hybridize to dsRNA isolated from North American strains of *C. parasitica*.[32] Furthermore, northern analysis performed on dsRNA from *C. parasitica* isolates from West Virginia, Michigan, France and Italy revealed several distinct homology groups.[33] dsRNA isolated from West Virginia strains did not cross-hybridize to dsRNA from Michigan *C. parasitica* strains, and none of the dsRNAs from North American strains tested cross-hybridized to dsRNA of French or Italian origin.[16,33] Recently, a *C. parasitica* isolate collected in New Jersey was found to contain dsRNA that cross-hybridized to dsRNA of European origin.[4] This was the first report of homology between dsRNA from North American and European isolates.

*b.   Molecular Organization of Hypoviruses
      from Europe*

Within the last 5 years, the basic genomic organization of the dsRNA associated with the French hypovirulent strain EP713 has been elucidated and represents the first designated *Cryphonectria* hypovirus (CHV1-713), a newly established family of viruses—the Hypoviridae.[34] The dsRNA from CHV1-713 can be grouped into three size classes, large (L), medium (M), and small(S). The L-dsRNA is a single molecule ~12.7 kb in length. The M-dsRNA size class consists of one or more molecules 8.0 – 10.0 kb in length, while S-dsRNAs consist of two or more bands which range in size from 0.6 – 1.7 kb.[35] The M- and S-dsRNA size classes result from internal deletions of the L-dsRNA.[36] The 3'-terminus of all dsRNA size classes is characterized by a stretch of polyadenylic acid (poly[A]).[35]

The EP713 L-dsRNA, which has been cloned as cDNA and sequenced, is 12,712 bp in length. The strand of L-dsRNA terminating with 3' poly(A) is the coding strand and contains two large open reading frames designated ORF A and ORF B.[37] The junction between ORF A and ORF B consists of a 1-bp overlap of the two ORFs, suggesting that a –1 frameshift may occur following the termination of ORF A and prior to the initiation of ORF B.[37] ORF A, which has been studied extensively, is 1869 nucleotides (nt) in length and encodes two polypeptides, p29 and p40.[38] The protein p29 is a protease, which is autocatalytically released from the ORF A polyprotein during translation.[38] The amino acid sequence of this protease is similar to that of the potyvirus helper component protease, known as HC-Pro.[39] A function for p40 has not been determined. Protoplasts of a virulent strain of *C. parasitica* were transformed with a cDNA copy of ORF A, which was found to confer some phenotypic traits associated with European hypovirulent isolates.[40] These traits included loss of pigmentation, suppression of sporulation, and a reduction in the accumulation of the fungal enzyme laccase. This suppression was mapped to p29 within ORF A, which was found to be necessary, but not fully responsible for these altered traits.[41] Although the biological function of laccase in *C. parasitica* is unknown, the enzyme is down-regulated in European dsRNA-containing hypovirulent isolates.[42-45] Surprisingly, these transformants were not reduced in virulence, demonstrating that the symptoms associated with the hypovirulent phenotype can be variable and segregated from the overall aggressiveness of the strain.

ORF B, which is ~9.5 kb in length, encodes a polyprotein that includes a protease domain (48 kDa) which contains amino acid motifs similar to those found in ORF A.[37] Within the deduced amino acid sequence of ORF B, two other putative domains have been identified, which would encode an RNA-dependent RNA polymerase and an RNA helicase.[46] It has been hypothesized that the dsRNA viruses of *C. parasitica* originated from ssRNA viruses.[37,46] since the putative domains were reported to be similar to those of ssRNA viruses, including barley yellow mosaic virus (BaYMV) which is a poty-like virus that is transmitted through a fungal vector.[46]

Choi and Nuss[45] transformed protoplasts of a virulent *C. parasitica* with a full-length cDNA copy of CHV1-713. Transformants were hypovirulent and possessed all of the phenotypic traits normally associated with fungal strains carrying CHV1-713. This was the first time genes encoded by dsRNA were conclusively demonstrated to be responsible for hypovirulence. Significantly, full-length dsRNA identical to CHV1-713 was isolated from the cytoplasm of transformants. These dsRNA molecules were produced from the cDNA copy, which had integrated into the chromosome of the fungus. These cytoplasmic dsRNAs were capable of cytoplasmic transfer via hyphal anastomosis to virulent strains, converting them to hypovirulence. The integrated copy of the viral cDNA has been shown to be mitotically and meiotically stable.[47] It is theorized that these genetically engineered hypovirulent strains should have a major advantage over naturally occurring *C. parasitica* isolates in biocontrol applications in that the cDNA copy of the dsRNA is integrated into the fungal genome. Sexual crosses between these strains and virulent strains yielded ascospore (sexual) progeny containing chromosomally integrated forms and cytoplasmic forms of CHV1-713 and were hypovirulent.[45] This is exceptional because, except for one specific case,[48] dsRNA has not been observed to be transmitted to ascospore progeny in naturally occurring hypovirulent strains.[22]

c.   Hypoviruses in North America

Hypovirulence is best known as a naturally occurring biocontrol responsible for the recovery of the chestnut tree and subsequently the many industries still utilizing the chestnut in Europe.[17,27] Not as well known is that many blighted, but surviving, small stands of American chestnut can be found within Michigan[5] and other localized areas in eastern North America including Ontario.[3] Most of Michigan is not within the natural range of American chestnut trees; however, trees can be found planted throughout the state.[49] In fact, all recovering chestnut stands within Michigan are located outside of the natural range.[17] Recovering stands contain hypovirulent strains and stands not recovering have few if any detectible hypovirulent strains.[49]

Obvious differences in culture morphology can be identified among different dsRNA-containing North American hypovirulent isolates. Hypovirulent isolates have various shades of pigmentation, uneven to smooth colony margins, and usually have slow to moderate growth rates when compared to virulent isolates.[3,5,50] In Michigan, characterization of the dsRNA revealed three dsRNA types based on sequence homology.[33,51] The predominant hypovirus in Michigan, designated CHV3-GH2, from *C. parasitica* strain GH2, contains three dsRNA segments of ~9.0, 3.5 and 0.8 kb in length. Many strains containing the CHV3-GH2-like dsRNA possess only the large, ~9.0-kb segment.[33] A second homology type consists of only two small dsRNA segments ~2.3 and 1.9 kb in length. This genome has only been identified in *C. parasitica* isolates from Roscommon, Michigan and was first isolated from strain RC1.[5] The third homology type is represented by a single isolate and contains a single dsRNA segment larger than the large segment of GH2 (>10 kb). While all three classes of dsRNA genomes are present in Michigan, natural mixed infections are generally not found in natural isolates. Mixed infections were shown to be possible, however, as CHV3-GH2 and dsRNA from RC1 were combined into the same nuclear background.[52,53] The co-infected strain was markedly reduced in virulence when compared to strains containing either virus genome alone. The viral genomes segregated independently in the asexual progeny, and new sizes of dsRNA segments were not observed in the progeny.[53]

The CHV3-GH2 hypovirus genome is made up of three segments of dsRNA (9.0, 3.5, and 0.8 kb), two with homology and one without homology to the others. The two larger segments cross-hybridize to each other, but not to the ~0.8-kb segment.[54] By sequence analysis of cDNA clones, it was determined that the ~0.8-kb segment does not produce protein products (D. Nuss, personal communication), and thus appears to be a satellite[55] genome of the larger segments. The significance of the satellite-like dsRNA segment is not known, but in light of the effects some satellites have on symptom expression, its role may be significant.[56] Further analysis of the ~9.0- and 3.5-kb segments of CHV3-GH2the GH2 dsRNA genome revealed that the 3.5-kb segment results from an internal deletion within the 9.0-kb segment, as both the 5′ and 3′ ends of the two segments are homologous.[54] The internal deletion appears to be responsible for many, if not all, dsRNA segments smaller than the largest segment in other hypoviruses studied.[36] The 9.0-kb segment of CHV3-GH2 has been partially sequenced, and appears to have a genomic organization similar to CHV1-713 in that a large ORF is present which may contain RNA-dependent RNA polymerase and RNA helicase domains.[57]

In New Jersey, hypovirulent *C. parasitica* strain NB58 contains a type of dsRNA that is found abundantly in that area.[50] The complete nucleotide sequence of NB58 dsRNA, recently designated hypovirus CHV2-NB58, has been characterized and found to be similar to CHV1-713 in size (12.5 kb) and organization.[58] The two viruses share approximately 60% nucleotide homology[58] and less than 40% homology with CHV3-GH2.[57] Two open reading frames designated ORF A and ORF B are connected through a pentanucleotide sequence UAAUG, which terminates ORF A and initiates ORF B. The ORF A protein product does not undergo autoproteolysis as does the protein product of ORF A of CHV1-713.[58] However, the ORF B products of the two hypoviruses were similar; the identified motifs were highly conserved.[58] In a survey of the dsRNAs associated with the hypovirulent *C. parasitica* population in New Jersey, polymerase chain reaction (PCR) products demonstrated that all New Jersey isolates were closely related and distinguishable from CHV1-713.[50] Further analysis of the PCR products indicated that none of the viruses in New Jersey were identical, suggesting genetic drift among dsRNA sequences within the New Jersey population. Re-isolation of a hypovirulent isolate from the same tree that yielded CHV2-NB58 4 years earlier produced identical sequences among the PCR products, suggesting that dsRNA sequences in a specific fungal thallus are conserved.[50]

d.   The Future of Hypoviruses in Biocontrol of Phytopathogenic Fungi

Hypovirulence in *C. parasitica* provides a unique biocontrol model for the establishment of transmissible forms of hypovirulence in other pathosystems. Since most phytopathogenic fungi are either uninfected or infected with benign forms of viruses, it would seem at the outset that hypovirulence could only be used in systems where it has been found to occur naturally. One of the most exciting prospects for biological control of fungi other than chestnut blight was recently reported by Chen et al.,[59] where synthetic transcripts of full-length CHV1-713 coding strands were found to be infectious when electroporated into fungal spheroplasts. In these experiments, the mycovirus transcript was used to expand fungal host range to include the phytopathogens *C. cubensis*, *C. havanensis*, and *E. gyrosa*. The viral transcript was used to expand viral host range because the cDNA-derived hypovirus RNA when transformed into fungi must undergo posttranslational splicing which other species of fungi cannot carry out. The synthetic viral transcript circumvents this step. The expansion of the host range to *E. gyrosa* suggested to Chen et al.[59] that virus-mediated hypovirulence will find broader application in biocontrol of alternative pathosystems.

B.   MITOCHONDRIAL-ASSOCIATED HYPOVIRULENCE IN CRYPHONECTRIA PARASITICA

1.   Cryphonectria parasitica

Transmissible hypovirulence associated with mitochondrial defects has been characterized in only two fungal pathogens, *C. parasitica*[60] and *O. ulmi*.[20] As discussed above, reduced virulence or aggressiveness is closely related to growth rate, which in turn is correlated with respiratory efficiency. Hypovirulent strains that show aberrant colony morphology, reduced growth, high levels of cyanide-resistant mitochondrial alternative oxidase (a symptom of respiratory "distress"), and maternal inheritance of hypovirulence and which are not infected with dsRNA viruses resemble the so-called senescent strains of *Neurospora* in which mitochondrial defects cause induction of cyanide-resistant respiration.[61-64] Cyanide-resistant respiration occurs in mitochondria of many fungi and higher plants which have a branched-chain, electron-transport system in which one branch is a conventional cytochrome system and the other is an alternative, cyanide-resistant pathway utilizing an alternative oxidase. The alternative pathway is not linked to oxidative phosphorylation and the function of alternative oxidase is not known in fungi. Those strains in which a high percentage of total respiratory oxygen consumption occurs through the alternative pathway usually show respiration impairment. Therefore, high levels of cyanide-resistant respiration in fungi can be a symptom of mitochondrial impairment. In dsRNA-free hypovirulent *C. parasitica* isolates recovered from nonlethal cankers on surviving chestnut trees in Michigan and Ontario, a large percentage of respiration is through the alternative pathway.[60] The mitochondrial hypovirulent phenotype and altered respiration are transmissible to virulent strains of *C. parasitica* in a manner similar to that of viral-mediated hypovirulence.[65] In experiments attempting to prove that mitochondrial impairment can lead to the hypovirulent phenotype in *C. parasitica*, mutants for high levels of alternative oxidase were induced in a normal virulent strain and linked to transmissible hypovirulence.[66,67] These experiments suggest an exciting new strategy for inducing hypovirulence in strains where viruses are not associated with hypovirulence.

dsRNAs have recently been isolated from the mitochondria of *C. parasitica* and their significance in the hypovirulent phenotypes is clearly different from that of dsRNA-free mitochondrial-associated hypovirulence.

2.   *Ophiostoma ulmi*

In hypovirulent strains of *O. ulmi*, dsRNA is found in the mitochondria as previously indicated (Section II.A.1, above). However, diseased isolates also contain head-to-tail concatemers of plasmid-like elements homologous to the mitochondrial DNA. Transfer of the diseased state to healthy strains did not result from the transfer of mtDNA, but did result in the formation of similar plasmid-like elements in the recipient strain. The connection between the presence of dsRNA, the plasmid DNA and the cytochrome deficiency is not well understood, but it is possible that mutant mitochondria could be involved in this form of transmissible hypovirulence.

III.   SUMMARY

The biological control of fungal diseases of plants will be an important aspect of biocontrol as fungicides continue to be tightly regulated. Innovative and unique opportunities exist for creative researchers willing to investigate both the molecular and ecological aspects of the disease triangle and the biocontrol system superimposed over the pathosystem. Fungal viruses which induce in their host a hypovirulence phenotype where natural host defense systems are capable of resisting infection offer hope for control. If viruses cannot be found in targeted pathogens, then it may be possible to construct viruses from other fungi, including saprophytic fungi.[15] It is possible that some of the benign viruses in saprophytic fungi, could, if moved into new fungal hosts, alter growth rates and induce hypovirulence.

Virulence of pathogens can also be correlated with respiration and mitochondria. Just as viruses can be transmitted from infected strains to uninfected strains, mitochondria or mtDNA are also capable of transmission to recipient strains, at least in the laboratory. Introducing (or inducing) abnormal mitochondria into otherwise healthy fungal strains would appear at first fruitless as the lowered fitness of the strain and/or rapid loss of defective mitochondria in the advancing thallus would purge the thallus of this disability. However, it has been shown that defective mitochondria not only can compete with the healthy mitochondria, but can replace them. The process that causes this displacement is not well understood and has been named "suppressiveness" because it is related to the disappearance of wild-type cytoplasmic factors from yeast zygotes generated by mating wild-type cells with certain cytoplasmic *petite* mutants.[63,64,68–70] Hypovirulence through suppressiveness theoretically could be induced in many fungi, including plant pathogens, and may offer a viable strategy for biological control.

REFERENCES

1. Farrand, S. K., *Agrobacterium radiobacter* strain K84: A model biocontrol system, in *New Directions in Biocontrol: Alternatives for Suppressing Agricultural Pests and Diseases*, Baker, R. R. and Dunn, P. E., Eds., Alan R. Liss, New York, 1990, 679.

2. Nuss, D. L., Biological control of chestnut blight: an example of virus-mediated attenuation of fungal pathogenesis, *Microbiol. Rev.*, 56, 561, 1992.

3. Dunn, M. M. and Boland, G. J., Hypovirulent isolates of *Cryphonectria parasitica* in southern Ontario, *Can. J. Plant Pathol.*, 15, 245, 1993.

4. Hillman, B. I., Fulbright, D. W., Nuss, D. L., and Van Alfen, N. K., Hypoviridae, in *Virus Taxonomy: Sixth Report of the International Committee for the Taxonomy of Viruses*, Murphy, F. A. et al., Eds., Springer-Verlag, New York, 1995, 261.

5. Fulbright, D. W., Weidlich, W. H., Haufler, K. Z., Thomas, C. S., and Paul, C. P., Chestnut blight and recovering American chestnut trees in Michigan, *Can. J. Bot.*, 61, 3164, 1983.

6. Brasier, C. M., A cytoplasmically transmitted disease of *Ceratocystis ulmi, Nature,* 305, 220, 1983.

7. Boland, G. J., Hypovirulence and double-stranded RNA in *Sclerotinia sclerotiorum, Can. J. Plant Pathol.*, 14, 10, 1992.

8. Naiki, T. and Cook, R. J., Factors in loss of pathogenicity in *Gaeumannomyces graminis* var. *tritici, Phytopathology*, 73, 1652, 1983.

9. Chumley, F. G. and Valent, B., Genetic analysis of melanin-deficient, nonpathogenic mutants of *Magnaporthe grisea, Mol. Plant Microb. Interact.*, 3, 135, 1990.

10. Buck, K. W., Fungal virology — an overview, in *Fungal Virology*, Buck, K.W., Ed., CRC Press, Boca Raton, FL, 1986, 1.

11. Hollings, M., Mycoviruses: viruses that infect fungi, *Adv. Virus Res.*, 22, 1978.

12. Ganesa, C., Chang, Y.-J., Flurkey, W. H., Rhandhawa, Z. I., and Bozarth, R. F., Purification and molecular properties of teh toxin coded by *Ustilago maydis* virus P4, *Biochem. Biophys. Res. Commun.*, 162, 651, 1989.

13. Tao, J., Ginsberg, I. Banerjee, N., Held, W., Koltin, Y., and Bruenn, J. A., *Ustilago maydis* KP6 killer toxin: structure, expression in *Saccharomyces cerevisiae*, and relationship to other cellular toxins, *Mol. Cell. Biol.*, 10, 1373, 1990.
14. Thomashow, L. S., Weller, D. M., Bonsall, R. F., and Peirson, L. S., Production of the antibiotic phenazine-1-carboxylic acid by fluorescent *Pseudomonas* species in the rhizosphere of wheat, *Appl. Environ. Microbiol.*, 56, 908, 1990.
15. Kim, J. W. and Bozarth, R. F., Intergeneric occurrence of related fungal viruses: the *Aspergillus ochraceous* virus complex and its relationship to the *Penicillium stoloniferum* virus S, *J. Gen. Virol.*, 66, 1991, 1985.
16. Enebak, S. A., MacDonald, W. L., and Hillman, B. I., Effect of dsRNA associated with isolates of *Cryphonectria parasitica* from the central Appalachians and their relatedness to other dsRNAs from North America and Europe, *Phytopathology*, 84, 528, 1994.
17. MacDonald, W. L. and Fulbright, D. W., Biological control of chestnut blight: use and limitations of transmissible hypovirulence, *Plant Disease*, 75, 656, 1991.
18. Rogers, H. J., Buck, K. W., and Brasier, C. M., Double-stranded RNA in diseased isolates of the aggressive subgroup of the Dutch elm fungus *Ophiostoma ulmi*, in *Viruses of Fungi and Simple Eukaryotes*, Koltin, Y. and Leibowitz, M., Eds., Marcel Dekker, New York, 1988, 327.
19. Brassier, C. M., *Ophiostoma novo-ulmi* sp. nov., causative agent of current Dutch elm disease pandemics, *Mycopathologia*, 115, 151, 1991.
20. Rogers, H. J., Buck, K. W., and Brasier, C. M., A mitochondrial target for double-stranded RNA in diseased isolates of the fungus that causes Dutch elm disease, *Nature*, 329, 558, 1987.
21. Brasier, C. M., The d-factor in *Ceratocystis ulmi* — its biological characteristics and implications for Dutch elm, in *Fungal Virology*, Buck, K. W., Ed., CRC Press, Boca Raton, FL, 1986, 177.
22. Anagnostakis, S. L., Biological control of chestnut blight, *Science*, 215, 466, 1982.
23. Grente, J., Les formes hypovirulentes d'*Endothia parasitica* et les espoirs de lutte contre le chancre du chataignier, *C.R. Hebd. Seances Acad. Agric. France*, 51, 1033, 1965.
24. Grente, J. and Berthelay-Sauret, S., Biological control of chestnut blight in France, in *Proc. Am. Chestnut Symp.*, MacDonald, W. L., Cech, F. C., Luchok, J., and Smith, H.C., Eds., West Virginia University Books, Morgantown, 1978, 30.
25. Day, P. R., Dodds, J. A., Elliston, J. E., Jaynes, R. A., and Anagnostakis, S. L., Double-stranded RNA in *Endothia parasitica*, *Phytopathology*, 67, 1393, 1977.
26. Elliston, J. E., Jaynes, R. A., Day, P. R., and Anagnostakis, S. L., A native American hypovirulent strain of *Endothia parasitica*, *Proc. Am. Phytopathol. Soc.*, 4, 83, 1977.
27. Heiniger, U. and Rigling, D., Biological control of chestnut blight in Europe, *Annu. Rev. Phytopathol.*, 32, 581, 1994.
28. Dodds, J. A., Association of type 1 viral-like dsRNA with club-shaped particles in hypovirulent strains of *Endothia parasitica*, *Virology*, 107, 1, 1980.
29. Newhouse, J. R. and MacDonald, W. L., Virus-like particles in hyphae and conidia of European hypovirulent (dsRNA-containing) strains of *Cryphonectria parasitica*, *Can. J. Bot.*, 68, 90, 1990.
30. Hansen, D. R., Van Alfen, N. K., Gillies, K., and Powell, W. A., Naked dsRNA associated with hypovirulence of *Endothia parasitica* is packaged in fungal vesicles, *J. Gen. Virol.*, 66, 2605, 1985.
31. Fahima, T., Kazmierczak, P., Hansen, D. R., Pfeiffer, P., and Van Alfen, N. K., Membrane-associated replication of an unencapsidated double-stranded RNA of the fungus, *Cryphonectria parasitica*, *Virology*, 195, 81, 1993.

32. L'Hostis, B., Hiremath, S. T., Rhoads, R. E., and Ghabrial, S. A., Lack of sequence homology between double-stranded RNA from European and American hypovirulent strains of *Endothia parasitica, J. Gen. Virol.,* 66, 351, 1985.

33. Paul, C. P. and Fulbright, D. W., Double-stranded RNA molecules from Michigan hypovirulent isolates of *Endothia parasitica* vary in size and sequence homology, *Phytopathology* 78, 751, 1988.

34. Hillman, B. I., Tian, Y., Bedker, P. J., and Brown, M. P., A North American hypovirulent isolate of the chestnut blight fungus with European isolate-related dsRNA, *J. Gen. Virol.,* 73, 681, 1992.

35. Hiremath, S., L'Hostis, B., Ghabrial, S. A., and Rhoads, R. E., Terminal structure of hypovirulence-associated dsRNAs in the chestnut blight fungus *Endothia parasitica, Nucleic Acids Res.,* 14, 9877, 1986.

36. Shapira, R., Choi, G. H., Hillman, B. I., and Nuss, D. L., The contribution of defective RNAs to the complexity of viral-encoded double-stranded RNA populations present in hypovirulent strains of the chestnut blight fungus *Cryphonectria parasitica, EMBO J.,* 10, 741, 1991b.

37. Shapira, R., Choi, G. H., and Nuss, D. L., Virus-like genetic organization and expression strategy for a double-stranded RNA genetic element associated with biological control of chestnut blight, *EMBO J.,* 10, 731, 1991a.

38. Choi, G. H., Shapira, R., and Nuss, D. L., Cotranslational autoproteolysis involved in gene expression from a double-stranded RNA genetic element associated with hypovirulence of the chestnut blight fungus, *Proc. Natl. Acad. Sci. U.S.A.,* 88, 1167, 1991a.

39. Choi, G. H., Pawlyk, D. M., and Nuss D. L., The autocatalytic protease p29 encoded by a hypovirulence-associated virus of the chestnut blight fungus resembles the potyvirus-encoded protease HC-Pro, *Virology,* 183, 747, 1991b.

40. Choi, G. H. and Nuss, D. L., A viral gene confers hypovirulence-associated traits to the chestnut blight fungus, *EMBO J.,* 11, 473, 1992a.

41. Craven, M. G., Pawlyk, D. M., Choi, G. H., and Nuss, D. L., Papain-like protease p29 as a symptom determinant encoded by a hypovirulence-associated virus of the chestnut blight fungus, *J. Virol.,* 67, 6513, 1993.

42. Rigling, D., Heiniger, U., and Hohl, G.H.R., Reduction of laccase activity in dsRNA-containing hypovirulent strains of *Cryphonectria parasitica, Phytopathology,* 79, 219, 1989.

43. Hillman, B. I., Shapira, R., and Nuss, D. L., Hypovirulence-associated suppression of host functions in *Cryphonectria parasitica* can be partially relieved by high light intensity, *Phytopathology,* 80, 950, 1990.

44. Rigling, D. and Van Alfen, N. K., Regulation of laccase biosynthesis in the plant-pathogenic fungus *Cryphonectria parasitica* by double-stranded RNA, *J. Bacteriol.,* 173, 8000, 1991.

45. Choi, G. H. and Nuss, D. L., Hypovirulence of the chestnut blight fungus conferred by an infectious viral cDNA, *Science,* 257, 800, 1992b.

46. Koonin, E. V., Choi, G. H., Nuss, D. L., Shapira, R., and Carrington, J. C., Evidence for common ancestry of a chestnut blight hypovirulence-associated double-stranded RNA and a group of positive-strand RNA plant viruses, *Proc. Natl. Acad. Sci. U.S.A.,* 88, 10647, 1991.

47. Chen, B., Choi, G. H., and Nuss, D. L., Mitotic stability and nuclear inheritance of integrated viral cDNA in engineered hypovirulent strains of the chestnut blight fungus, *EMBO J.,* 12, 2991, 1993.

48. Polashock, J. J. and Hillman, B. I., A small mitochondrial double-stranded (ds) RNA element associated with a hypovirulent strain of the chestnut blight fungus and ancestrally related to yeast cytoplasmic T and W dsRNAs, *Proc. Natl. Acad. Sci. U.S.A.*, 91, 8680, 1994.

49. Fulbright, D. W., Paul, C. P., and Garrod, S. W., Hypovirulence: a natural control of chestnut blight, in *Biocontrol of Plant Diseases*, Vol. 2, Mukeiju, K. G. and Garg, K. L., Eds., CRC Press, Boca Raton, FL, 1988, 122.

50. Chung, P., Bedker, P. J., and Hillman, B. I., Diversity of *Cryphonectria parasitica* hypovirulence-associated double-stranded RNAs within a chestnut population in New Jersey, *Phytopathology*, 84, 984, 1994.

51. Fulbright, D. W., Molecular basis for hypovirulence and its ecological relationships, in *New Directions in Biocontrol: Alternatives for Suppressing Agricultural Pests and Diseases*, Baker, R. R. and Dunn, P. E., Eds., Alan R. Liss, New York, 1990, 693.

52. Vannini, A., Smart, C. D., and Fulbright, D. W., The comparison of oxalic acid production in vivo and *in vitro* by virulent and hypovirulent *Cryphonectria (Endothia) parasitica*, *Physiol. Mol. Plant Path.*, 43, 443, 1993.

53. Smart, C. D. and Fulbright, D. W., Characterization of a strain of *Cryphonectria parasitica* doubly-infected with hypovirulence-associated dsRNA viruses, *Phytopathology*, 85, 491, 1994.

54. Tartaglia, J., Paul, C. P., Fulbright, D. W., and Nuss, D. L., Structural properties of double-stranded RNAs associated with biological control of chestnut blight fungus, *Proc. Natl. Acad. Sci. U.S.A.*, 83, 9109, 1986.

55. Collmer, C. W. and Howell, S. H., Role of satellite RNA in the expression of symptoms caused by plant viruses, *Annu. Rev. Phytopathol.*, 30, 419, 1992.

56. Rodriguez-Alvarado, G., Kurath, G., and Dodds, J. A., Symptom modification by satellite tobacco mosaic virus in pepper types and cultivars infected with helper tobamoviruses, *Phytopathology*, 84, 617, 1994.

57. Durbahn, C. M., Molecular Characterization of dsRNA-Associated Hypovirulence in Michigan Isolates of *Cryphonectria parasitica*, Ph.D. dissertation, Michigan State University, East Lansing, 1992, chap. 3.

58. Hillman, B. I., Halpern, B. T., and Brown, M. P., A viral dsRNA element of the chestnut blight fungus with a distinct genetic organization, *Virology*, 201, 241, 1994.

59. Chen, B., Choi, G. H., and Nuss, D. L., Attenuation of fungal virulence by synthetic infectious hypovirus transcripts, *Science,* 264, 1762, 1994.

60. Mahanti, N., Bertrand, H., Monteiro-Vitorello, C., and Fulbright, D. W., Elevated mitochondrial alternative oxidase activity in dsRNA-free, hypovirulent isolates of *Cryphonectria parasitica*, *Physiol. Mol. Plant Pathol.*, 42, 455, 1993.

61. Lambowitz, A. M. and Slayman, C., Cyanide-resistant respiration in *Neurospora crassa*, *J. Bacteriol.*, 108, 1087, 1971.

62. Lambowitz, A. M. and Zannoni, D., Cyanide-insensitive respiration in *Neurospora*. Genetic and biophysical approaches, in *Plant Mitochondria*, Ducet G. and Lance, C., Eds., Elsevier, North-Holland Biomedical Press, Amsterdam, 1978, 283.

63. Bertrand, H., Collins, R. A., Stohl, L. L., Goewert R., and Lambowitz, A. M., Deletion mutants of *Neurospora crassa* mitochondrial DNA and their relationship to the "stop-start" growth phenotype, *Proc. Natl. Acad. Sci. U.S.A.*, 77, 6032, 1980.

64. Griffiths, A. J. F., Fungal senescence, *Annu. Rev. Gen.*, 26, 351, 1992.

65. Mahanti, N. and Fulbright, D. W., Cytoplasmic transfer of mitochondria associated with alternate oxidase and hypovirulence in *Cryphonectria parasitica*, *Phytopathology*, 81, 1139, 1991.

66. Vitorello, C., Monteiro, Mahanti, N., Bell, J. A., Fulbright, D. W., and Bertrand, H., Cytoplasmic factors other than dsRNA viruses may determine hypovirulence in some strains of *Cryphonectria parasitica* in *Program of the Seventh Fungal Genetics Conference*, Asilomar, CA, 1993, 27.

67. Monteiro-Vitorello, C. B., Bell, J. A., Fulbright, D. W., and Bertrand, H., A cytoplasmically-transmissible hypovirulence phenotype associated with mitochondrial DNA mutations in the chestnut blight fungus *Cryphonectria parasitica, Proc. Natl. Acad. Sci. U.S.A.*, submitted.

68. Ephrussi, B., Margerie-Hottinguer, H. de, and Roman, H., Suppressiveness: a new factor in the genetic determinism of the synthesis of respiratory enzymes in yeast, *Proc. Natl. Acad. Sci. U.S.A.*, 41, 1065, 1955.

69. Bertrand, H., Argan, C. A., and Szakacs, N. A., Genetic control of the biogenesis of cyanide insensitive respiration in *Neurospora crassa*, in *Mitochondria 1983*, Schweyen, R. J., Wolf, K., and Kaudewitz F., Eds., Walter deGruyter, New York, 1983, 495.

70. Bertrand, H., Griffiths, A. J. F., Court, D. A., and Cheng, C. K., An extrachromosomal plasmid is the etiological precursor of kalDNA insertion sequences in the mitochondrial chromosome of senescent *Neurospora, Cell*, 47, 829, 1986.

5 The Role of Molecular Biology in Developing Biological Controls for Plant Parasitic Nematodes

*Graham R. Stirling, Lois Eden,
and Elizabeth Aitken*

TABLE OF CONTENTS

I. INTRODUCTION

Plant parasitic nematodes are ubiquitous in their distribution but because of their small size and cryptic habitat, their economic importance is frequently underestimated. Nevertheless, nematode population densities reach damaging levels in many

0-8493-2442-4/96/$0.00+$.50
© 1996 by CRC Press, Inc.

agricultural cropping systems and the problems they initiate cause losses worth billions of dollars to the world's food and fiber crops.[1] Effective controls have not been developed for all crops, but losses from nematodes are sometimes reduced through strategies such as crop rotation and cultivar resistance. Chemical nematicides are also widely used, particularly in intensively managed horticultural and ornamental crops, but their future is uncertain because of health and environmental concerns[2] and community demands for more environmentally friendly pest and disease control practices.

Plant parasitic nematodes spend most, if not all, of their lives in soil, where they co-exist with a wide range of other soil organisms. Since some of these organisms parasitise or prey on nematodes, or have the capacity to influence nematode behaviour, they are therefore of interest to those involved in the development of biological control strategies for nematodes. A broad overview of the parasites and predators of nematodes is available in several texts,[3-5] but the most important antagonists are the nematophagous fungi, obligately parasitic bacteria in the genus *Pasteuria*, predatory mites, predacious nematodes, and a number of miscellaneous organisms (mainly bacteria) whose mechanisms of action are still to be determined. Many of these biological control agents, particularly the nematode-trapping and egg-parasitic fungi and the bacterium *Pasteuria penetrans* (Thorne) Sayre & Starr, have been tested for activity against nematodes in small-scale tests in the laboratory, glasshouse, and field,[5] but biological control currently plays no more than a limited role as a practical control strategy for nematodes.

The factors which have limited the development of biological control have received considerable attention in recent years and there is some consensus that our lack of understanding of the variability of biological control agents, our inadequate knowledge of their mechanisms of action, and our inability to adequately monitor organisms following their introduction into soil is limiting progress. However, recent developments in molecular biology and biotechnology provide today's researcher with the techniques required to tackle such issues. Molecular studies aimed at understanding the plant–nematode relationship,[6-8] identifying resistance genes in plants,[9,10] inserting novel genes into plants to confer nematode resistance,[8,11,12] and identifying nematodes at species and subspecies level[13] are, for example, already having an impact on nematode control strategies involving host plant resistance, and similar progress could be expected if molecular technologies were applied to other areas of plant nematology. This review will therefore examine the role that molecular biology might play in removing some of the impediments to progress in biological control.

II.   RESOLUTION OF TAXONOMIC PROBLEMS

Many natural enemies of nematodes are difficult to identify. They are relatively obscure organisms of uncertain taxonomic position whose descriptions are often inadequate because an emphasis on morphology limits the availability of taxonomic characters. There have been relatively few studies of variability within taxa or of relationships between taxa, and few taxonomic monographs are available to aid species identification. This absence of reliable taxonomic information means that it is difficult to study potentially useful biological control agents. Traditional taxonomy can fill some of the gaps in our knowledge, but modern techniques in molecular biology could also be usefully employed to resolve some of these taxonomic issues.

A. IMPROVING SPECIES IDENTIFICATION AND UNDERSTANDING PHYLOGENY

The difficulties involved in identifying the nematophagous fungi are typical of the problems involved identifying all with biological control agents of nematodes. The nematode-trapping fungi, for example, consist of numerous species within a number of superficially similar genera of Deuteromycetes (Fungi Imperfecti), namely, *Arthrobotrys*, *Geniculifera*, *Duddingtonia*, *Dactylella*, and *Monacrosporium*. Genera are separated largely on characteristics of conidial formation, while species differentiation relies on conidial morphology. However, variation in key characteristics can occur following repeated transfer on agar,[14] and this variability, when combined with the multiplicity of species and the limited number of morphological characters available, means that species within this group are difficult to identify using traditional methods.

Molecular studies can help overcome taxonomic problems, but they have been rarely used with the nematophagous fungi. One of the few examples involves the nematode egg parasite *Verticillium chlamydosporium* Goddard. This species, which is usually regarded as a species complex,[15–17] was recently subdivided into two species on the basis of physiological and morphological characteristics.[18] Assessment of this complex using restriction fragment-length polymorphism (RFLP) analysis did not support this subdivision but did indicate that *V. chlamydosporium* was distinct from other non-nematophagous *Verticillium* spp.[19]

In addition to providing practical means of identifying organisms, taxonomists strive to develop systems of classification which infer evolutionary relationships among organisms. Traditional taxonomy is based on phenotypic characters, but these do not necessarily reveal underlying genetic homologies because they may have arisen separately due to convergent evolution. The morphology of organisms is programmed by information stored in their DNA so that classifications based on genome structure and nucleotide sequence homology give more insight into evolutionary relationships. The techniques of molecular systematics have already been used widely to improve our knowledge of the phylogeny of bacteria[20] and fungi,[21] and there is no doubt they could also be used to improve the classification of nematophagous microorganisms.

One example where phylogenetic analysis would be useful involves the *Pasteuria penetrans* group of bacteria. Three nematode-parasitic species of *Pasteuria* have been recognised on the basis of host range and morphological features such as the shape of endospores, vegetative colonies, and sporangia.[22,23] However, there are many subspecies and pathotypes within each "species" and there is a need to define the boundaries between populations that may infect as many as 175 different nematode species. Also, *P. penetrans* can be cultured only in its nematode host[24] or in association with host cells,[25] and phylogenetic analysis could provide information which might aid the development of *in vitro* culture techniques. The morphology of its vegetative stages and heat-stable endospores suggest that *P. penetrans* is related to actinomycetes and bacilli, respectively. Attempts to cultivate the organism have therefore been based on the assumption that its requirements for growth are similar to these groups of bacteria. The chances of successful *in vitro* culture would perhaps be improved if the closest relatives of *P. penetrans* were known.

A study such as that carried out recently with the host-specific bacteria *Rhizobium* and *Bradyrhizobium* might be particularly useful with *P. penetrans*. Random amplified polymorphic DNA (RAPD) analysis of 84 isolates of these genera showed that most of them produced unique DNA fingerprints. It was possible to cluster the isolates into several specific groups[26] and at the same time, to differentiate strains within biovars of *R. leguminosarum* (Frank) Frank.

There are many examples in the literature of the use of molecular techniques to understand the phylogeny of plant pathogenic fungi, and similar methodologies could be used with the nematophagous fungi. One useful technique involves sequence analysis of nonconserved regions of rDNA. When this technique was used recently to compare three species of *Puccinia* rust, the internal transcribed spacer (ITS) region showed sufficient variability to enable species differentiation and phylogenetic analysis.[27] A similar study of Sclerotiniaceae showed that species with morphologically distinct stromata also were distinct on the basis of ITS analysis.[28]

B.   ASSESSING VARIABILITY WITHIN SPECIES

One common problem faced by researchers in biological control is the selection of isolates with good biocontrol potential from the vast array of isolates that are available. For example, in species such as *V. chlamydosporium* where variability has been studied, isolates differ considerably in their growth at different temperatures, their ability to produce chlamydospores, their rhizosphere competence, and their virulence.[16,29] Molecular studies are useful in this situation, because random analysis of DNA can produce patterns which associate with particular characters. These patterns (or markers) may be due to genetic linkage to specific characters or they may reflect evolutionary changes. Nevertheless, if biological control activity is found to be linked to the presence of certain markers, the markers can be used in screening programs to identify new isolates with biocontrol potential. Such studies could also be employed to separate nematode parasitic species from those attacking plants (e.g., *Verticillium*) or humans (e.g., *Paecilomyces*), or to help characterise a specific biocontrol isolate for which patent protection might be required.

Although there have been few molecular studies of variation within nematophagous organisms, such studies have been attempted with both entomogenous and plant pathogenic fungi and bacteria. RFLP analysis of a virulent and less virulent mutant of *Beauvaria bassiana* (Balsamo) Vuillemin using the ß-tubulin gene of *Neurospora crassa* Shear & Dodge as the probe showed that there were differences between the two strains.[30] In another study using RFLP analysis, serotypes of 14 mosquitocidal strains of *Bacillus sphaericus* Meyer & Neide isolated from 3 genera of mosquito, were distinguished from each other. A nonpathogenic strain had a different DNA marker pattern from pathogenic strains.[31]

Within the plant pathogenic fungi, both RFLP and RAPD analyses have been used to relate pathogenicity to groupings that were established according to molecular analysis. RFLP mapping of *Magnaporthe grisea* (Hebert) Barr, a fungus causing rice blast disease, has shown unique patterns for different pathotypes and revealed a family of repeated sequences.[32] It also suggested a relationship between the number of copies of these sequences in the genome and pathogenicity[33] and indicated a region of the genome possibly involved in pathogenicity.[34] RAPD analysis of pathogenic and nonpathogenic races of the fungus *Cochliobolus carbonum* Nelson also produced different patterns for two of the four pathogenic races and for the nonpathogenic race studied.[35] In contrast, RAPD analysis of 115 isolates of *Puccinia striiformis* Westend f.sp. *tritici*, which causes stripe rust of wheat, showed a very low association between the patterns produced and pathogenicity.[36] These examples demonstrate that useful information on variability can be obtained when the areas amplified and the probes chosen correspond to the parts of the genome that control the trait being studied.

III.   UNDERSTANDING HOST SPECIFICITY

Organisms with the capacity to parasitise or prey on nematodes exhibit considerable diversity in the specificity of their relationship with nematodes. At one end of the nutritional spectrum, some of the egg-parasitic and nematode-trapping fungi and the omnivorous invertebrates are facultative parasites and predators, so that they are not dependent on nematodes for their nutrition. At the other end of the spectrum, species such as the bacterium *P. penetrans* and the endoparasitic fungus *Hirsutella rhosilliensis* Minter & Brady are nutritionally dependent on nematodes. Species in this latter group have considerable potential as biological control agents, but there is a need to develop a better understanding of their host–parasite relationships.

The relationship between *P. penetrans* and *Meloidogyne* spp. is perhaps the most well studied of the host–parasite systems involving nematodes. A number of studies have shown that *P. penetrans* spores attach to individuals in some populations of *Meloidogyne* but not to others. Since spore attachment is the first phase of the infection process, this is an indication of variability in virulence within the bacterium.[37–40] However, in all of these studies, the taxonomy of both host and pathogen was based largely on morphological criteria which were not sufficiently precise to classify the organisms at the subspecific level. Interactions between host and parasite are determined at that level and, in recent years, there has been considerable improvement in our understanding of such interactions. Studies using polyclonal antibodies raised against endospores[41] and radioiodination of nematode cuticular proteins[42] have shown the nature and quantity of proteins on the endospore and on the nematode cuticle are involved in host specificity. The receptor on the nematode cuticle has since been found to be a glycoprotein, with the sugar moiety probably acting as the receptor.[43] Since these chemical configurations are determined at the genetic level, a gene for gene relationship similar to that which exists between bacterial plant pathogens and their hosts[44] may be involved in spore attachment. For each gene that conditions a particular reaction in the host, there may be a corresponding gene in the bacterium that conditions attachment. Such interactions are amenable to study using molecular techniques.

In soil, there are numerous predator–prey relationships between nematodes, mites, and other soil fauna but very little is known about the feeding preferences of nematode predators. One way of studying food preferences would be to retrieve predators from soil and analyse their gut contents. However, it may be difficult to identify specific DNA sequences in gut contents because the DNA would be quickly digested by enzymes in the gut. Assays with monoclonal antibodies are more likely to be useful and have been used to detect the protein remains of one slug species, *Arion hotensis* Férussac, from a background of nineteen other nonmollusc species in the gut of the beetle *Abax parallelepipedus* Piller & Mitterpacher.[45] In a similar study, whitefly (*Bemisia tabaci* Gennadius) remains were detected using monoclonal antibodies in beetle (*Geocoris punctipes* Say) gut contents.[46]

IV. MONITORING INTRODUCED BIOLOGICAL CONTROL AGENTS

One of the most basic problems confronting researchers working with biological control systems in soil is the difficulty of separating introduced biological control agents from resident species and then monitoring the occurrence and activity of the introduced species over time. Biological control experiments with nematodes have often produced inconsistent results and this inconsistency will only be overcome by developing a better understanding of the behaviour of organisms following their introduction into soil. We need to know how far biological control agents move from the point of introduction; whether they are able to colonise the root or soil niche occupied by the target nematode; and whether they are in a resting or metabolically active state. We also need to know whether their activity is affected by environmental factors such as moisture and temperature; by biological factors such as changes in the soil microflora or nematode density; or by farming practices such as cultivation and agrochemical usage.

The traditional methods which have been used to monitor nematophagous fungi and bacteria in soil have a number of limitations. Dilution plating onto selective media is commonly used for *V. chlamydosporium* and *Paecilomyces lilacinus* (Thom) Samson,[47–50] but the media are only partially selective and are not effective in all soils. Predatory and endoparasitic nematophagous fungi can be enumerated using most-probable number techniques,[51,52] and bioassay methods are used to detect spores of parasites with a more obligate nature such as *Hirsutella rhossiliensis* and *P. penetrans*.[53,54] However, such procedures are tedious and time consuming and improvements in tracking methods are urgently needed. Molecular techniques have been used to monitor bacteria and fungi in other environments and could possibly be used to monitor antagonists of nematodes in soil.

Molecular detection and enumeration techniques for microorganisms can be broadly subdivided into those that do or do not involve culture of the organism. In one group of methods, culture of the microorganism incorporates or is followed by either phenotypic expression or genotypic detection involving nucleic acid hybridisation. Nonculturable procedures involve extracting nucleic acids from environmental samples followed by polymerase chain reaction (PCR) and/or hybridisation to detect a unique sequence, or direct cell counts using immunological and hybridisation techniques incorporating fluorescent or colorimetric markers.[55]

A.   DETECTION USING REPORTER GENES

A reporter gene (or a genetic marker) is a gene that is present in an organism, usually as a result of mutation or genetic transformation. The gene confers on the organism a property it would not normally have, such as antibiotic resistance. The standard genetic markers for bacteria are resistance genes which allow the organism to grow on a medium containing an antibiotic such as rifampicin or tetracycline[56] or nutritional reporter genes that allow the organism to use unusual food sources. The *lac Z* gene, for example, encodes for ß-galactosidase, which allows bacteria to utilise lactose. The presence of ß-galactosidase is easily monitored by incorporation of the indicator substrate X-gal into the culture medium, as this is converted into a blue coloured product which is easily visualised.[57] Reporter genes have been introduced into filamentous fungi which allow them to grow in the presence of the fungicide benomyl,[58] the antibiotic hygromycin B,[59] and on acetamide and acrylamide.[60] However, before such markers can be utilised, microorganisms must be culturable on artificial media. These markers would be useful for many nematophagous fungi and some bacteria, but obligate parasites and bacteria in a viable but nonculturable state could not be detected using such methods.

The GUS system, which is used widely in transgenic plants as a reporter gene, encodes the enzyme ß-glucuronidase from *Escherichia coli* (Migula) Castellani & Chalmers that cleaves a glucuronide substrate to produce a colorigenic product.[61] It has been used successfully in bacteria and in filamentous fungi, and transformed organisms can be detected readily after culture on artificial media. They could also be detected *in situ* if the masking effects of soil could be overcome. The GUS gene is a useful marker for studies of root colonisation, as the colour change can be observed directly on the roots. The spatial distribution of the microorganisms on the roots can therefore be determined.[62]

Luciferase reporter genes from bacteria and fireflies have been used in plants, bacteria, and yeasts but there have not yet been any widely published reports of its use in filamentous fungi. When the gene is present and luciferin is added in the presence of oxygen and ATP, light is emitted and can be detected using autophotography or optical fibre light measurement techniques.[63,64] As with the GUS reporter gene, luciferase is useful for studying root colonisation, as transformed microorganisms can be observed after culture on artificial media or observed directly on the roots.

Despite recent progress in the use of genetic markers for monitoring microorganisms, genetic markers may be unstable and can be lost when the transformed microorganism is added to soil.[65] Instability tends to be a particular problem with plasmid-borne markers, and greater stability can be achieved when markers are integrated into genomic DNA. Stable fungal transformants have been selected following regrowth of colonies from single conidia and subsequent growth without selection pressure.[60] Another problem with genetic markers is that their presence may affect competitive ability. Marked strains are sometimes unable to out-compete their parent strains,[65-67] whereas, in other cases, markers persist and do not affect competitiveness.[68-70] Also, the use of markers may not always be the most appropriate technology because of governmental restrictions and community concern regarding the release of genetically modified microorganisms into the environment.

B.   DETECTION USING DNA PROBES

There have been major advances in nucleic acid hybridisation technology in recent years and it is now used routinely and effectively for environmental monitoring in disciplines as diverse as food microbiology, forensics, human diagnostics, and plant pathology.[71] The foundation of this technology is a labelled DNA probe which hybridises specifically to the specimen's DNA at regions which share homologous stretches of complementary nucleotides. These probes can be used to detect specific sequences of DNA introduced into microorganisms or to detect already existing unique sequences.[55,56] However, when DNA is introduced into a microorganism as a specific target for a probe, the problems of marker stability and environmental fitness of genetically modified organisms (previously discussed) have to be overcome.

Molecular probes can be used with either culturable or unculturable bacteria. If an organism can be cultured, colonies are transferred from the culture medium to a membrane filter where the cells are lysed, and the probe is added and allowed to hybridise with the target DNA.[55] Such a procedure has been used in fresh water to detect *Pseudomonas putida* (Trevison) Migula containing an added fragment of eukaryotic DNA.[69]

Although nucleic acid hybridisation technology has revolutionised the manner in which microorganisms are monitored in many environments, there are constraints to its use in soil. Humic polymers in soil are detrimental to DNA and few extraction protocols yield DNA in a form that can be restricted, cloned, or PCR-amplified.[56,72] These problems have been overcome to a certain extent with methods that reduce interference from humic compounds,[73,74] but further improvements will be required before this technology can be routinely applied in soil. One possibility is to utilise 16s rRNA probes to detect DNA from microorganisms without the need to extract them from the soil. The labelled probes are combined with epifluorescence microscopy or with flow cytometry and have been used to detect 14 different *Filobacter* strains in soil.[75,76]

Another aspect that requires further research is the use of molecular technologies to quantify microorganisms in soil. At present, methods suitable for detecting organisms are available but quantification methodologies require further development. Such methodologies are likely to be based on PCR, as it is a very powerful quantitative tool when appropriate optimisation and validation procedures are used.[77] One potentially useful improvement is competitive PCR, in which a known DNA fragment is added to the amplification reaction to permit estimation of reaction efficiency. This parameter is then used in an equation which estimates initial DNA concentration.[78,79]

V.   DETERMINING MECHANISMS OF ACTION

Parasitism, predation, competition, and antibiosis are key components of many biological control systems and most interactions between antagonists and nematodes involve one of these processes. However, most studies of these interactions have been relatively superficial, so that our knowledge of the mechanisms of action of most biological agents is limited. Hypotheses about mechanisms have been proposed but often there is little concrete evidence that particular mechanisms are actually involved. For example, egg-parasitic fungi such as *Dactylella oviparasitica* Stirling & Mankau, *Paecilomyces lilacinus*, and *Verticillium chlamydosporium* are chitinolytic, leading to speculation that chitinases are involved in the infection process.[80,81] Proteases and other lytic enzymes have also been implicated in egg parasitism,[82,83] while ultrastructural studies suggest that toxic metabolites from fungi may prevent eggs from hatching and may incapacitate juveniles.[84,85] The production of nematicidal substances in culture has also been presented as evidence that toxins play a role in pathogenicity.[86] Obviously, evidence for the involvement of enzymes and toxins in pathogenicity is largely circumstantial and more direct evidence is needed.

Molecular techniques have now been used to obtain such evidence for several other biological control systems. In one of the first examples, a phenazine antibiotic produced by a strain of *Pseudomonas fluorescens* Migula was shown to be involved in the suppression of the fungal pathogen *Gaeumannomyces graminis* (Sacc.) von Arx & Oliver var. *tritici* Walker.[87] Transposon Tn5 mutagenesis was used to generate mutants defective in phenazine synthesis, and these mutants failed to suppress the pathogen, whereas the phenazine-producing, wild-type strain was suppressive. When the mutants were complemented with a DNA clone which codes for phenazine from a library of the wild-type genome, both antibiotic synthesis and fungal suppression were coordinately restored. Tn5 mutants have been used by others to show that fluorescent-siderophore production by *P. fluorescens* did not account for all its antagonistic activity against *Pythium ultimum* Trow,[88] and to demonstrate that in the absence of its flagella, the plant growth-stimulating bacterium *P. fluorescens* was impaired in its ability to colonise plant roots.[89] Such studies demonstrate the precision of techniques in molecular genetics and provide an indication of how these techniques might be used to determine the role of toxins and enzymes in the pathogenicity of the egg-parasitic fungi.

There are many other biological control agents of nematodes where similar molecular techniques could be applied. Several recent studies have yielded rhizobacteria[90-94] and actinomycetes[95] with activity against *Heterodera* and *Meloidogyne*, but their mechanisms of action are not known. Production of ammonia or other nematicidal compounds or interference with host finding processes by blockage of receptors on roots or by modification of root exudates have been suggested as possible mechanisms. All these hypotheses are amenable to examination using molecular techniques.

One area where molecular techniques could also be usefully employed is in understanding the predacious behaviour of the nematode trapping fungi. Attraction of nematodes to chemicals produced by the trapping structures, induction of traps by chemicals associated with nematodes, and immobilisation of prey by nematoxins may be involved in the trapping process,[3] but there is no more than observational evidence to support such hypotheses. Molecular studies could provide direct evidence of the involvement of particular chemicals in each of these processes, a knowledge of the identity of the chemicals and an understanding of the genetics of their production and regulation. Such studies are therefore essential if we are to make attempts to improve the predacious activity of these fungi.

VI. MANIPULATION OF USEFUL GENES

To date, all of the work on bacterial and fungal biological control agents of plant-parasitic nematodes has been done with wild-type strains. However, this situation may soon change because methods are now available to manipulate the genomes of both prokaryotic and eukaryotic cells so that existing traits of an organism can be improved or new traits added. Possible improvements could include enhancing the production of an enzyme or toxin responsible for biological control activity, extending an organism's host range, adding a new mechanism of action, improving rhizosphere competence, or increasing tolerance to pesticides.

One of the constraints to the use of genetic manipulation to improve biological control activity is the lack of information on the mechanisms of action of many biological control agents of nematodes. Genes responsible for production of chitinase and collagenase are two possible candidates for manipulation, as these two enzymes are often implicated in biological control activity. Chitin and collagen are the major constituents of the nematode egg shell and cuticle, respectively,[96,97] and nematodes can be controlled by using these substrates to stimulate chitinolytic and collagenolytic activity in soil.[98-102] Biological control agents that act by producing chitinase or collagenase could perhaps be improved by manipulating the organism so that it produces excess enzyme. This could be achieved by increasing the number of copies of the gene responsible for activity, by altering its promoter so that the gene product is produced constitutively rather than by induction, or by replacing its promoter region with a more active one. Alternatively, the genes could be transferred to an organism where the production of these enzymes would complement other useful characteristics.

There are several recent examples of the transformation of chitinolytic soil microorganisms in an attempt to improve biological control activity against chitin-containing fungal plant pathogens. Provided chitinase and collagenase are shown to be detrimental to nematodes, similar techniques could be applied with antagonists of nematodes. *Rhizobium meliloti* Dangeard was transformed with the *Serratia marcescens* Bizio chitinase gene to produce transformants with plant growth promoting qualities similar to the wild type. The nodules produced chitinase and the cell-free extract of the nodules lysed *Rhizoctonia solani* Kühn.[103] *Trichoderma harzianum* Rifai was also transformed with the *S. marcescens* chitinase gene under the control of the 35S-constitutive promoter of cauliflower mosaic virus. Transformants produced chitinase constitutively and, when tested for antagonistic activity against *Sclerotium rolfsii* Sacc. in dual culture, lytic activity was greater than the wild-type strain.[60,104] However, improved biological control activity was not demonstrated.

Although the techniques needed to transfer chitinase and collagenase from one microorganism to another are available, there are problems with the application of this technology that require consideration. The environmental fitness of *Cochliobolus heterostrophus* (Drechs.) Drechs. was reduced after transformation with constitutively expressed genes,[105] suggesting that the energy required in the constitutive production of a product may diminish the competitive ability of transformed organisms. Constitutively produced chitinase may also have detrimental effects on beneficial, nontarget organisms such as mycorrhizal fungi. An additional complicating factor is that different chitinases have different effects on organisms,[60] so that chitinases which act specifically against nematode chitin may have to be selected. Such a step may reduce the impact of the transformant on other soil organisms, but specificity towards nematodes could perhaps be obtained more readily by targeting components of nematodes other than the egg shell or cuticle. Gene products which target materials produced by sensory organs or oesophageal glands and interfere with host-finding or feeding are, for example, more likely to be specific to plant-parasitic nematodes. Genes encoding for these products could, for example, be inserted into microorganisms that normally reside in the rhizosphere or on the nematode's cuticle.

Since toxin production is thought to be responsible for biological activity of some nematophagous organisms, nematicidal toxin genes could be manipulated for strain improvement in the same manner as has been discussed for enzymes. The progress made in improving the activity of the biological insecticide *Bacillus thuringiensis* Berliner (*B.t.*) gives some indication of what can be achieved. Mass screening of *B.t.* has resulted in the discovery of a large variety of toxins with a wide range of specificities and genes for many of these have been isolated, sequenced and analysed.[106] Most of the toxin genes are plasmid-borne and, in natural populations, these plasmids are often transferred from one strain to related strains. Researchers have therefore been able to produce strains with several toxin genes.[107] This is an advantage because the availability of a wide range of toxin products minimises opportunities for the development of resistance. Toxin genes have also been modified to yield toxins with higher specific activity, thereby reducing the quantity required.

Because of the instability of *B.t.*-based products in the environment, molecular biology has been used to develop other delivery systems.[107] *P. fluorescens* has been transformed with toxin plasmids, and these cells are mass produced and then killed by a chemical fixative to prevent rupture of the cell walls. This process prevents the premature release of the toxin from the *B.t.* endospore due to lysis, as the toxin is not released until cells are applied to soil. Genetically engineered endophytic bacteria and transgenic plants have also been produced with insecticidal activity. The endophytic bacteria inhabit the xylem of plants and the toxin is ingested by insects when they feed on the plant. The advantage of this over transgenic plants is that the plant does not have to expend metabolic energy on producing the toxin constitutively. Also, the toxin can be changed from season to season in order to minimise the risk of insect resistance. The recent observation that some strains of *B. thuringiensis* are effective against plant-parasitic nematodes[108] raises the possibility that these methodologies could be used to improve efficacy against nematodes.

Since nematophagous fungi must act in situations where fungicides are commonly used, fungicide resistance would be a desirable attribute to add to these biological control agents. Benomyl resistance has been obtained in fungi by chemical mutagenesis,[109] and many attempts have been made to transform into fungi the ß-tubulin gene from *Neurospora crassa* that confers this resistance. Most of these attempts have been made with *Trichoderma* spp. and have involved transformation of protoplasts or protoplast fusion. Although the percentage of successful transformants from either of these methods is very low, some genetically stable benomyl-resistant isolates have been produced. A few of these transformants have also shown improved rhizosphere competence[110] or improved biological control ability.[110,111] As techniques for genetically manipulating filamentous fungi improve, it will be possible to apply this technology to nematophagous fungi.

Another trait which is useful in any parasite or predator of nematodes is rhizosphere competence. Potentially useful biocontrol agents must have this characteristic, as plant-parasitic nematodes are closely associated with plant roots during the feeding process and rhizosphere competence enables them to survive in the same environment as the target nematode. The basis of rhizosphere competence is not yet understood, but is probably a complex phenomenon involving many genes. Since it is difficult to genetically engineer complex traits into a biological control agent, it has been suggested that more success would be achieved by introducing biocontrol properties into an already rhizosphere competent microorganism.[60,112] Plant growth-promoting bacteria such as *Rhizobium* and *Pseudomonas fluorescens* are two potentially useful recipients of biocontrol attributes.

It is obvious from the preceding discussion that gene transfer technologies offer many opportunities to enhance biological control activity. To date, however, they have produced few practical outcomes in any field of biological control. The techniques required to genetically modify microorgansims have been available for more than a decade, but regulatory restrictions on the release of such organisms have tended to stifle progress. Gene transfer is common in bacteria,[113] while fungi have transposons and cytoplasmically transmitted genetic elements which may be transmitted horizontally to other fungi,[114] so that there is concern that genes introduced into bacterial and fungal biological control agents may spread to native microorganisms. If genetic manipulation is to play more than a limited role in the future of biological control, these concerns will have to be addressed. Public perceptions of genetically engineered microorganisms will also have to improve. Recent work[115] has assessed the variability of 25 isolates of *Hirsutella rhossiliensis* with RAPD analysis using eight primers. There was substantial genetic variation but it was not associated with pathogenicity.

VI. CONCLUDING REMARKS

Although the techniques of molecular biology are now widely used in the biological sciences, they have not been adopted to any great extent by those interested in developing biological controls for nematodes. Much of the material presented in this review is therefore speculative and many of the opportunities for research have been illustrated by examples from other disciplines. Hopefully, this situation will change in the future as more resources are devoted to biological control. Molecular techniques are likely to find application in many areas, but perhaps they could be most usefully employed in improving our understanding of the mode of action of biological control agents. Once the biochemical and genetic processes involved in pathogenesis are known, it should be possible to use this information to make biological control more successful.

REFERENCES

1. Sasser, J. N. and Freckman, D. W., A world perspective on nematology: the role of the society, in *Vistas on Nematology*, Veech, J. A. and Dickson, D. W., Eds., Society of Nematologists, Hyattsville, MD, 1987, 7.
2. Thomason, I. J., Challenges facing nematology: environmental risks with nematicides and the need for new approaches, in *Vistas on Nematology*, Veech, J. A. and Dickson, D. W., Eds., Society of Nematologists, Hyattsville, MD, 1987, 469.
3. Barron, G. L., *The Nematode Destroying Fungi*, Canadian Biological Publications, Guelph, 1977.
4. Poinar, G. O. J. and Jansson, H.-B., *Diseases of Nematodes*. Vol. 1 and 2, CRC Press, Boca Raton, FL, 1988.
5. Stirling, G. R., *Biological Control of Plant Parasitic Nematodes*, C.A.B International, Wallingford, CT, 1991.
6. Burrows, P. R., Molecular analysis of the interactions between cyst nematodes and their hosts, *J. Nematol.*, 24, 338, 1992.
7. Goddijn, O. J. M., Lindsey, K., van der Lee, F. M., Klap, J. C., and Sijmons, P. C., Differential gene expression in nematode-induced feeding structures of transgenic plants harbouring promoter-gusA fusion constructs, *Plant J.*, 4, 863, 1993.
8. Hyman, B. C. and Opperman, C. H., Contemporary approaches to the study of host-parasite interactions: an introduction, *J. Nematol.*, 24, 321, 1992.
9. Pineda, O., Bonierbale, M. W., Plaisted, R. L., Brodie, B. B., and Tanksley, S. D., Identification of RFLP markers linked to the *H1* gene conferring resistance to the potato cyst nematode *Globodera rostochiensis*, *Genome*, 36, 152, 1993.

10. Williamson, V. M., Ho, J.-Y., and Ma, H. M., Molecular transfer of nematode resistance genes, *J. Nematol.*, 24, 234, 1992.

11. Burrows, P. R. and Jones, M. G. K., Cellular and molecular approaches to the control of plant parasitic nematodes, in *Plant Parasitic Nematodes in Temperate Agriculture*, Evans, K., Trudgill, D. L., and Webster, J. M., Eds., CAB International, Wallingford, CT, 1993, 609.

12. Opperman, C. H., Taylor, C. G., and Conkling, M. A., Root-knot nematode-directed expression of a plant root-specific gene, *Science*, 263, 221, 1994.

13. Curran, J. and Robinson, M. P., Molecular aids to nematode diagnosis, in *Plant Parasitic Nematodes in Temperate Agriculture*, Evans, K., Trudgill, D. L., and Webster, J. M., Eds., C.A.B. International, Wallingford, CT, 1993, 545.

14. van Oorschot, C. A. N., Taxonomy of the *Dactylaria* complex, V. A review of *Arthrobotrys* and allied genera, *Stud. Mycol.*, 26, 61, 1985.

15. Bursnall, L. A. and Tribe, H. T., Fungal parasitism in cysts of *Heterodera*. II. Egg parasites of *H. schachtii*, *Trans. Br. Mycol. Soc.*, 62, 595, 1974.

16. Irving, F. and Kerry, B. R., Variation between strains of the nematophagous fungus, *Verticillium chlamydosporium* Goddard. II. Factors affecting parasitism of cyst nematode eggs, *Nematologica*, 32, 474, 1986.

17. Kerry, B. R., Irving, F., and Hornsey, J. C., Variation between strains of the nematophagous fungus, *Verticillium chlamydosporium* Goddard. I. Factors affecting growth *in vitro*, *Nematologica*, 32, 461, 1986.

18. Gams, W., A contribution to the knowledge of nematophagous species of *Verticillium*, *Neth. J. Plant Pathol.*, 94, 123, 1988.

19. Carder, J. H., Segers, R., Butt, T. M., Barbara, D. J., von Mende, N., and Coosemans, J., Taxonomy of the nematophagous fungi *Verticillium chlamydosporium* and *V. suchlasporium* based on secreted enzyme activities and RFLP analysis, *J. Invertebr. Pathol.*, 62, 178, 1993.

20. Woese, C. R., Bacterial evolution, *Microbiol. Rev.*, 51, 221, 1987.

21. Bruns, T. D., White, T. J., and Taylor, J. W., Fungal molecular systematics, *Annu. Rev. Ecol. Syst.*, 22, 525, 1991.

22. Sayre, R. M., Wergin, W. P., Nishizawa, T., and Starr, M. P., Light and electron microscopical study of a bacterial parasite from the cyst nematode, *Heterodera glycines*, *J. Helminthol. Soc. Wash.*, 58, 69, 1991.

23. Starr, M. P. and Sayre, R. M., *Pasteuria thornei* sp. nov. and *Pasteuria penetrans* sensu stricto emend., mycelial and endospore-forming bacteria parasitic, respectively, on plant parasitic nematodes of the genera *Pratylenchus* and *Meloidogyne*, *Ann. Inst. Pasteur/Microbiol.*, 139, 11, 1988.

24. Stirling, G. R. and Wachtel, M. F., Mass production of *Bacillus penetrans* for the biological control of root knot nematodes, *Nematologica*, 26, 308, 1980.

25. Previc, E. P. and Cox, R. J., U.S. Patent 5094954, 1992, Production of endospores from *Pasteuria penetrans* by culturing with explanted tissues from nematodes.

26. Dooley, J. J., Harrison, S. P., Mytton, L. R., Dye, M., Cresswell, A., Skot, L., and Beeching, J. R., Phylogenetic grouping and identification of *Rhizobium* isolates on the basis of random amplified polymorphic DNA profiles, *Can. J. Microbiol.*, 39, 665, 1993.

27. Zambino, P. J. and Szabo, L. J., Phylogenetic relationships of selected cereal and grass rusts based on rDNA sequence analysis, *Mycologia*, 85, 401, 1993.

28. Carbone, I. and Kohn, L. M., Ribosomal DNA sequence divergence within internal transcribed spacer 1 of the Sclerotiniaceae, *Mycologia*, 85, 415, 1993.

29. Kerry, B. R., Simon, A., and Rovira, A. D., Observations on the introduction of *Verticillium chlamydosporium* and other parasitic fungi into soil for control of the cereal cyst-nematode *Heterodera avenae*, *Ann. Appl. Biol.*, 105, 509, 1984.
30. Kosir, J. M., MacPherson, J. M., and Khachatourians, G. G., Genomic analysis of a virulent and a less virulent strain of the entomopathogenic fungus *Beauveria bassiana*, using restriction fragment length polymorphisms, *Can. J. Microbiol.*, 37, 534, 1991.
31. Abadjieva, A., Miteva, V., and Grigorova, R., Genomic variations in mosquitocidal strains of B*acillus sphaericus* detected by M13 DNA fingerprinting, *J. Invertebr. Pathol.*, 60, 5, 1992.
32. Levy, M., Romao, J., Marchetti, M. A., and Hamer, J. E., DNA fingerprinting with a dispersed repeated sequence resolves pathotype diversity in the rice blast fungus, *Plant Cell*, 3, 95, 1991.
33. Hamer, J. E., Molecular probes for rice blast disease, *Science*, 252, 632, 1991.
34. Hamer, J. E. and Givan, S., Genetic mapping with dispersed repeated sequences in the rice blast fungus: mapping the *SMO* locus, *Mol. Gen. Genet.*, 223, 487, 1990.
35. Jones, M. J. and Dunkle, L. D., Analysis of *Cochliobolus carbonum* races by PCR amplification with arbitrary and gene-specific primers, *Phytopathology*, 83, 366, 1993.
36. Chen, X. M., Line, R. F., and Leung, H., Relationship between virulence variation and DNA polymorphism in *Puccinia striiformis*, *Phytopathology*, 83, 1489, 1993.
37. Spaull, V. W., Observations on *Bacillus penetrans* infecting *Meloidogyne* in sugar cane fields in South Africa, *Rev. Nematol.*, 7, 277, 1984.
38. Stirling, G. R., Host specificity of *Pasteuria penetrans* within the genus *Meloidogyne*, *Nematologica*, 31, 203, 1985.
39. Sell, P. and Hansen, C., Beziehungen zwischen Wurzelgallen-Nematoden und ihrem naturlichen Gegenspieler *Pasteuria penetrans*, *Meded. Fac. Landbouwwet. Rijksuniv. Gent*, 52, 607, 1987.
40. Davies, K. G., Kerry, B. R., and Flynn, C. A., Observations on the pathogenicity of *Pasteuria penetrans*, a parasite of root-knot nematodes, *Ann. Appl. Biol.*, 112, 491, 1988.
41. Davies, K. G. and Danks, C., Interspecific differences in the nematode surface coat between *Meloidogyne incognita* and *M. arenaria* related to the adhesion of the bacterium *Pasteuria penetrans*, *Parasitology*, 105, 475, 1992.
42. Robinson, M. P., Delgado, J., and Parkhouse, R. M. E., Characterization of stage-specific cuticular proteins of *Meloidogyne incognita* by radio-iodination, *Physiol. Mol. Plant Pathol.*, 35, 135, 1989.
43. Davies, K. G., *In vitro* recognition of a 190 kDa putative attachment receptor from the cuticle of *Meloidogyne javanica* by *Pasteuria penetrans* spore extract, *Biocontrol Sci. Technol.*, 4, 367, 1994.
44. Bent, A., Carland, F., Dahlbeck, D., Innes, R., Kearney, B., Ronald, P., Roy, M., Salmeron, J., Whalen, M., and Staskawicz, B., Gene-for-gene relationships specifying disease resistance in plant-bacterial interactions, in *Advances in Molecular Genetics of Plant-Microbe Interactions*, Vol. 1, Hennecke, H. and Verma, D. P. S., Eds., Kluwer Academic Publishers, Dordrecht, 1991, 32.
45. Symondson, W. O. C. and Liddell, J. E., A monoclonal antibody for the detection of arionid slug remains in carabid predators, *Biol. Control*, 3, 207, 1993.
46. Hagler, J. R., Brower, A. G., Tu, Z., Byrne, D. N., Bradley-Dunlop, D., and Enriquez, F. J., Development of a monoclonal antibody to detect predation of the sweetpotato whitefly, *Bemisia tabaci*, *Entomol. Exp. Appl.*, 68, 231, 1993.
47. Mitchell, D. J., Kannwischer-Mitchell, M. E., and Dickson, D. W., A semi-selective medium for the isolation of *Paecilomyces lilacinus* from soil, *J. Nematol.*, 19, 255, 1987.

48. Cabanillas, E. and Barker, K. R., Impact of *Paecilomyces lilacinus* inoculum level and application time on control of *Meloidogyne incognita* on tomato, *J. Nematol.*, 21, 115, 1989.

49. Gaspard, J. T., Jaffee, B. A., and Ferris, H., *Meloidogyne incognita* survival in soil infested with *Paecilomyces lilacinus* and *Verticillium chlamydosporium*, *J. Nematol.*, 22, 176, 1990.

50. de Leij, A. A. M. and Kerry, B. R., The nematophagous fungus *Verticillium chlamydosporium* as a potential biological control agent for *Meloidogyne arenaria*, *Rev. Nematol.*, 14, 157, 1991.

51. Stirling, G. R., McKenry, M. V., and Mankau, R., Biological control of root knot nematodes (*Meloidogyne* spp) on peach, *Phytopathology*, 69, 806, 1979.

52. Dackman, C., Olsson, S., Jansson, H.-B., Lundgren, B., and Nordbring-Hertz, B., Quantification of predatory and endoparasitic nematophagous fungi in soil, *Microb. Ecol.*, 13, 89, 1987.

53. Stirling, G. R., Biological control of *Meloidogyne javanica* with *Bacillus penetrans*, *Phytopathology*, 74, 55, 1984.

54. McInnis, T. M. and Jaffee, B. A., An assay for *Hirsutella rhossiliensis* spores and the importance of phialides for nematode inoculation, *J. Nematol.*, 21, 229, 1989.

55. Saint, C. P. and McClure, N. C., Molecular developments for the tracking of microorganisms in the environment, *Australas. Biotechnol.*, 3, 76, 1993.

56. Kluepfel, D. A., The behavior and tracking of bacteria in the rhizosphere, *Annu. Rev. Phytopathol.*, 31, 441, 1993.

57. Maniatis, T., Fritsch, E. F., and Sambrook, J., *Molecular Cloning, a Laboratory Manual*, Cold Spring Harbor Laboratory, Cold Spring Harbor, NY, 1982.

58. Koenraadt, H. and Jones, A. L., Resistance to benomyl conferred by mutations in codon 198 or 200 of the beta-tubulin gene of *Neurospora crassa* and sensitivity to diethofencarb conferred by codon 198, *Phytopathology*, 83, 850, 1993.

59. Sivan, A., Stasz, T. E., Hemmat, M., Hayes, C. K., and Harman, G. E., Transformation of *Trichoderma* spp. with plasmids conferring hygromycin B resistance, *Mycologia*, 84, 687, 1992.

60. Chet, I., Barak, Z., and Oppenheim, A., Genetic engineering of microorganisms for improved biocontrol activity, in *Biotechnology in Plant Disease Control*, Chet, I., Ed., Wiley-Liss, New York, 1993, 211.

61. Jefferson, R. A., Burgess, S. M., and Hirsch, D., ß-Glucuronidase from *Escherichia coli* as a gene-fusion marker, *Proc. Nat. Acad. Sci. U.S.A.*, 83, 8447, 1986.

62. Vande Broek, A., Michiels, J., Van Gool, A., and Vanderleyden, J., Spatial-temporal colonization patterns of *Azospirillum brasilense* on the wheat root surface and expression of the bacterial *nifH* gene during association, *Mol. Plant-Microbe Interact.*, 6, 592, 1993.

63. Amin-Hanjani, S., Meikle, A., Glover, L. A., Prosser, J. I., and Killham, K., Plasmid and chromosomally encoded luminescence marker systems for detection of *Pseudomonas fluorescens* in soil, *Mol. Ecol.*, 2, 47, 1993.

64. de Weger, L. A., Dunbar, P., Mahafee, W. F., Lugtenberg, B. J. J., and Sayler, G. S., Use of bioluminescence markers to detect *Pseudomonas* spp. in the rhizosphere, *Appl. Environ. Microbiol.*, 57, 3641, 1991.

65. van Elsas, J. D., Trevors, J. T., van Overbeek, L. S., and Starodub, M. E., Survival of *Pseudomonas fluorescens* containing plasmids RP4 or pRK2501 and plasmid stability after introduction into two soils of different texture, *Can. J. Microbiol.*, 35, 951, 1989.

66. van Elsas, J. D., van Overbeek, L. S., and Fouchier, R., A specific marker, *pat*, for studying the fate of introduced bacteria and their DNA in soil using a combination of detection techniques, *Plant Soil*, 138, 49, 1991.

67. Halverson, L. J., Clayton, M. K., and Handelsman, J., Variable stability of antibiotic-resistance markers in *Bacillus cereus* UW85 in the soybean rhizosphere in the field, *Mol. Ecol.*, 2, 65, 1993.

68. Fredrickson, J. K., Bentjen, S. A., Bolton, H. Jr., Li, S. W., and van Voris, P., Fate of Tn5 mutants of root growth-inhibiting *Pseudomonas* sp in intact soil-core microcosms, *Can. J. Microbiol.*, 35, 867, 1989.

69. Genthner, F. J., Campbell, R. P., and Pritchard, P. H., Use of a novel plasmid to monitor the fate of a gentically engineered *Pseudomonas putida* strain, *Mol. Ecol.*, 1, 137, 1992.

70. Lindström, K., Lipsanen, P., and Kaijalainen, S., Stability of markers used for identification of two *Rhizobium galegae* inoculant strains after five years in the field, *Appl. Environ. Microbiol.*, 56, 444, 1990.

71. Miller, S. A. and Joaquim, T. R., Diagnostic techniques for plant pathogens, in *Biotechnology in Plant Disease Control*, Chet, I., Ed., Wiley-Liss, New York, 1993, 321.

72. Trevors, J. T., DNA extraction from soil, *Microb. Releases*, 1, 3, 1992.

73. Jacobsen, C. S. and Rasmussen, O. F., Development and application of a new method to extract bacterial DNA from soil based on separation of bacteria from soil with cation-exchange resin, *Appl. Environ. Microbiol.*, 58, 2458, 1992.

74. Tsai, Y.-L. and Olson, B. H., Rapid method for separation of bacterial DNA from humic substances in sediments for polymerase chain reaction, *Appl. Environ. Microbiol.*, 58, 2292, 1992.

75. Amann, R. I., Binder, B. J., Olson, R. J., Chisholm, S. W., Devereux, R., and Stahl, D. A., Combination of 16S rRNA-targeted oligonucleotide probes with flow cytometry for analyzing mixed microbial populations, *Appl. Environ. Microbiol.*, 56, 1919, 1990.

76. Amann, R. I., Krumholz, L., and Stahl, D. A., Fluorescent-oligonucleotide probing of whole cells for determinative, phylogenetic and environmental studies in microbiology, *J. Bacteriol.*, 172, 762, 1990.

77. Ferre, F., Quantitative or semi-quantitative PCR: reality versus myth, *PCR Methods Appl.*, 2, 1, 1992.

78. Gilliland, G., Perrin, S., Blanchard, K., and Bunn, H. F., Analysis of cytokine mRNA and DNA: Detection and quantitation by competitive polymerase chain reaction, *Proc. Nat. Acad. Sci. U.S.A.*, 87, 2725, 1990.

79. Buck, K. J., Harris, R. A., and Sikela, J. M., A general method for quantitative PCR analysis of mRNA levels for members of gene families: application to GABA_A receptor subunits, *BioTechniques*, 11, 636, 1991.

80. Stirling, G. R. and Mankau, R., Mode of parasitism of *Meloidogyne* and other nematode eggs by *Dactylella oviparasitica*, *J. Nematol.*, 11, 282, 1979.

81. Godoy, G., Rodríguez-Kábána, R., and Morgan-Jones, G., Parasitism of eggs of *Heterodera glycines* and *Meloidogyne arenaria* by fungi isolated from cysts of *H. glycines*, *Nematropica*, 12, 111, 1982.

82. Dunn, M. T., Sayre, R. M., Carrell, A., and Wergin, W. P., Colonization of nematode eggs by *Paecilomyces lilacinus* (Thom) Samson as observed with scanning electron microscope, *Scanning Electron Micros.*, 3, 1351, 1982.

83. Lopez-Llorca, L. V., Purification and properties of extracellular proteases produced by the nematophagous fungus *Verticillium suchlasporium*, *Can. J. Microbiol.*, 36, 830, 1990.

84. Morgan-Jones, G., White, J. F., and Rodríguez-Kábána, R., Phytonematode pathology: ultrastructural studies. I. Parasitism of *Meloidogyne arenaria* eggs by *Verticillium chlamydosporium*, *Nematropica*, 13, 245, 1983.

85. Morgan-Jones, G., White, J. F., and Rodríguez-Kábána, R., Phytonematode pathology: ultrastructural studies. II. Parasitism of *Meloidogyne arenaria* eggs and larvae by *Paecilomyces lilacinus*, *Nematropica*, 14, 57, 1984.

86. Cayrol, J.-C., Djian, C., and Pijarowski, L., Study of the nematocidal properties of the culture filtrate of the nematophagous fungus *Paecilomyces lilacinus*, *Rev. Nematol.*, 12, 331, 1989.

87. Thomashow, L. S. and Weller, D. M., Role of a phenazine antibiotic from *Pseudomonas fluorescens* in biological control of *Gaeumannomyces graminis* var *tritici*, *J. Bacteriol.*, 170, 3499, 1988.

88. Loper, J. E., Role of fluorescent siderophore production in biological control of *Pythium ultimum* by a *Pseudomonas fluorescens* strain, *Phytopathology*, 78, 166, 1988.

89. De Weger, L. A., van der Vlugt, C. I. M., Wijfjes, A. H. M., Bakker, P. A. H. M., Schippers, B., and Lugtenberg, B., Flagella of a plant growth stimulating *Pseudomonas fluorescens* strain are required for colonization of potato roots, *J. Bacteriol.*, 169, 2769, 1987.

90. Zavaleta-Mejia, E., The effect of soil bacteria on *Meloidogyne incognita* (Kofoid & White) Chitwood infection, *Diss. Abstr. Int.*, 46, 1018-B, 1985.

91. Becker, J. O., Zavaleta-Mejia, E., Colbert, S. F., Schroth, M. N., Weinhold, A. R., Hancock, J. G., and Van Gundy, S. D., Effects of rhizobacteria on root-knot nematodes and gall formation, *Phytopathology*, 78, 1466, 1988.

92. Oostendorp, M. and Sikora, R. A., Seed treatment with antagonistic rhizobacteria for the suppression of *Heterodera schachtii* early root infection of sugar beet, *Rev. Nematol.*, 12, 77, 1989.

93. Oostendorp, M. and Sikora, R. A., *In vitro* interrelationships between rhizosphere bacteria and *Heterodera schachtii*, *Rev. Nematol.*, 13, 269, 1990.

94. Oka, Y., Chet, I., and Spiegel, Y., Control of the rootknot nematode *Meloidogyne javanica* by *Bacillus cereus*, *Biocontrol Sci. Technol.*, 3, 115, 1993.

95. Dicklow, M. B., Acosta, N., and Zuckerman, B. M., A novel *Streptomyces* species for controlling plant-parasitic nematodes, *J. Chem. Ecol.*, 19, 159, 1993.

96. Perry, R. N. and Trett, M. W., Ultrastructure of the egg shell of *Heterodera schachtii* and *H. glycines* (Nematoda: Tylenchida), *Rev. Nematol.*, 9, 399, 1986.

97. Reddigari, S. R., Jansna, P. C., Premadrandram, D., and Hussey, R. S., Cuticular collagenous proteins of second-stage juveniles and adult females of *Meloidogyne incognita*; Isolation and partial characterization, *J. Nematol.*, 18, 294, 1986.

98. Mankau, R. and Das, S., The influence of chitin amendments on *Meloidogyne incognita*, *J. Nematol.*, 1, 15, 1969.

99. Spiegel, Y., Cohn, E., and Chet, I., Use of chitin for controlling plant parasitic nematodes. I. Direct effects on nematode reproduction and plant performance, *Plant Soil*, 95, 87, 1986.

100. Spiegel, Y., Chet, I., and Cohn, E., Use of chitin for controlling plant parasitic nematodes. II. Mode of action, *Plant Soil*, 98, 337, 1987.

101. Spiegel, Y., Chet, I., Cohn, E., Galper, S., and Sharon, E., Use of chitin for controlling plant parasitic nematodes. III. Influence of temperature on nematicidal effect, mineralization and microbial population buildup, *Plant Soil*, 109, 251, 1988.

102. Galper, S., Cohn, E., Spiegel, Y., and Chet, I., Nematicidal effect of collagen-amended soil and the influence of protease and collagenase, *Rev. Nematol.*, 13, 67, 1990.

103. Sitrit, Y., Barak, Z., Kapulnik, Y., Oppenheim, A. B., and Chet, I., Expression of *Serratia marcescens* chitinase gene in *Rhizobium meliloti* during symbiosis on alfalfa roots, *Mol. Plant-Microbe Interact.*, 6, 293, 1993.

104. Haran, S., Schickler, H., Pe'er, S., Logemann, S., Oppenheim, A., and Chet, I., Increased constitutive chitinase activity in transformed *Trichoderma harzianum*, *Biol. Control*, 3, 101, 1993.

105. Keller, N. P., Bergstrom, G. C., and Yoder, O. C., Effects of genetic transformation on fitness of *Cochliobolus heterostrophus*, *Phytopathology*, 80, 1166, 1990.

106. Lal, R. and Lal, S., *Genetic Engineering of Plants for Crop Improvement*, CRC Press, Boca Raton, FL, 1993.

107. Feitelson, J. S., Payne, J., and Kim, L., *Bacillus thuringiensis*: insects and beyond, *Bio/technology*, 10, 271, 1992.

108. Zuckerman, B. M., Dicklow, M. B., and Acosta, N., A strain of *Bacillus thuringiensis* for the control of plant-parasitic nematodes, *Biocontrol Sci. Technol.*, 3, 41, 1993.

109. Ahmad, J. S. and Baker, R., Rhizosphere competence of benomyl-tolerant mutants of *Trichoderma* spp, *Can. J. Microbiol.*, 34, 694, 1988.

110. Hayes, C. K., Improvement of *Trichoderma* and *Gliocladium* by genetic manipulation, in *Biological Control of Plant Diseases*, Tjamos, E. S., Papavizas, G. C., and Cook, R. J., Eds., Plenum Press, New York, 1992, 277.

111. Armstrong, J. L. and Harris, D. L., Biased DNA integration in *Colletotrichum gloeosporioides* f. sp. *aeschynomene* transformants with benomyl resistance, *Phytopathology*, 83, 328, 1993.

112. Juhnke, M. E., Mathre, D. E., and Sands, D. C., Identification and characterization of rhizosphere-competent bacteria of wheat, *Appl. Environ. Microbiol.*, 53, 2793, 1987.

113. Stewart, G. J., Natural transformation and its potential for gene transfer in the environment, in *Microbial Ecology: Principles, Methods, and Applications*, Levin, M. A., Seidler, R. J., and Rogul, M., Eds., McGraw-Hill, New York, 1992.

114. Kistler, H. C. and Miao, V. P. W., New modes of genetic change in filamentous fungi, *Annu. Rev. Phytopathol.*, 30, 131, 1992.

115. Tedford, E. C., Jaffee, B. A., and Muldoon, A. E., Variability among isolates of the nematophagous fungus *Hirsutella rhossiliensis*, *Mycol. Res.*, 98, 1127, 1994.

6 Genetically Enhanced Baculovirus Insecticides

H. Alan Wood

TABLE OF CONTENTS

I. INTRODUCTION

Baculoviruses are DNA viruses and comprise the largest and most widely studied group of viruses that are pathogenic to insects. The majority of baculoviruses have been reported from species in the order Lepidoptera, many of which are major agricultural or forest pests.[1]

For over a hundred years, naturally occurring baculovirus epizootics have been recognized as having the capacity in nature to regulate insect pest populations. Based on these observations, biological control programs have attempted to create artificial viral epizootics prior to the development of pest populations above economic thresholds.[2] It was considered that these viruses could serve as safe and effective alternatives to chemical insecticides. This approach has proven very successful in Brazil for the control of *Anticarsia gemmatalis* in over 2 million acres of soybean fields, using the *A. gemmatalis* nuclear polyhedrosis virus (AgMNPV).[3]

However, the use of baculoviruses as alternatives to chemical insecticides has generally been problematic for several reasons. First and foremost, the production costs of viral pesticides in most countries have been significantly higher than cost of chemical pesticides. Recent advances in the development of insect tissue culture bioreactors and serum-free media have significantly reduced in *vitro* production costs.[4] In addition, Crop Genetics International (Hanover, MD)[5] and AgriVirion (New York, NY)[6] have recently developed more cost-effective in *vivo* production systems. It is anticipated that these and other new technological advances will soon significantly reduce the production costs of viral pesticides.

Another deterrent to the commercialization of viral pesticides has been their limited market size. Studies with natural viral pesticides indicate that most baculoviruses have rather narrow host ranges.[7] For this reason, the commercial development of viral pesticides has been restricted to major crops. It should be kept in mind, however, that insecticides with activities against a limited number of host species are agronomically advantageous in that nontarget and beneficial insect populations are not affected.

A third problem associated with viral pesticides is their short half-life. Viral pesticides are inactivated by the sun's ultraviolet light. Accordingly, application and formulation technologies must be developed to optimize the efficacy of viral pesticides.

Another problem with the commercial use of baculovirus pesticides has been their slow speed of action. It can take from 5–15 days post-infection before larval death. Based on this slow speed of action relative to most chemical insecticides, viral pesticides have been considered to have limited commercial efficacy and have not generally been included in integrated pest management systems.

With the advent of molecular biology has come an interest in the development of recombinant viral pesticides which have more commercially attractive attributes than the naturally occurring viruses. These technologies have the potential to reduce production costs, extend host ranges and, most important, enhance the pesticidal properties of viral pesticides. Using the baculovirus expression vector technology,[8,9] foreign pesticidal genes can be easily inserted into baculoviruses. When these foreign gene products are expressed during virus replication, they can decrease the time to death or time to cessation of feeding.

Because of the availability of susceptible tissue culture systems, the insertion and expression of pesticidal genes has been quite simple to achieve with viruses such as the *Autographa californica* and *Bombyx mori* nuclear polyhedrosis viruses

(AcMNPV and BmNPV, respectively). It should be recognized, however, that tissue culture systems are not available for most other baculoviruses, and this has made the construction of recombinant virus isolates problematic.

To date, however, several pesticidal genes have been inserted into AcMNPV and BmNPV and evaluated under laboratory conditions for their improved pesticidal properties. In most studies, a polyhedrin gene was replaced with the foreign gene. Typically, the foreign genes have been placed under the transcriptional control of the very late polyhedrin or p10 baculovirus gene promoters. The following is a brief discussion of several of these novel recombinant constructs.

II. PESTICIDAL GENE CONSTRUCTS

A. DIURETIC HORMONE

Diuretic and antidiuretic hormones play important roles with insects in the regulation of water balance. They are used to control water secretion and resorption in response to environmental changes. Accordingly, it was considered that the expression of either hormone at high concentrations would disrupt the natural water balance of an insect larva, leading to death or cessation of feeding.

A diuretic hormone (DH) was isolated from *Manduca sexta*, and its amino acid sequence determined.[10] Based on this sequence and the codon usages for the polyhedrin gene of BmNPV, a synthetic gene was constructed.[11] Because the specific activity of the hormone was increased following amidation, a glycine residue was added to the C terminus. To facilitate the movement of the DH from the site of production to site of action, the signal sequence of the CP2 cuticle protein of *Drosophila melanogaster*[12] was added to the 5' end of the DH coding region. A DH expression vector was constructed by replacement of the BmNPV polyhedrin gene with the DH gene construct.[11]

Following injection of silkworm larvae with the DH-recombinant virus, the larvae had a 30% reduction in hemolymph volume as compared to uninfected and wild-type virus-infected larvae. The DH-recombinant virus infections of fifth-instar larvae also exhibited a small reduction in the LT_{50} (time to death of 50% of infected larvae) as compared to wild-type virus infections.

B. JUVENILE HORMONE ESTERASE

Juvenile hormone (JH) is a component of the larval endocrine system involved in metamorphosis. Early in the last larval instar, there is an increase in juvenile hormone esterase (JHE) which reduces the amount of JH. The absence of JH is necessary for the release of prothoracicotropic hormone which leads to a cessation of feeding and the initiation of metamorphosis to the pupal stage. Sparks and Hammock[13] have shown that inhibition of JHE activity results in a delay in metamorphosis and the development of giant larvae. Therefore, it was expected that increased JHE activity might cause a cessation in larval feeding and premature pupation.

Hammock et al.[14] inserted the JHE gene from *Heliothis virescens* into the AcMNPV genome under the control of the polyhedrin gene. A secretory leader sequence was included and more than 90% of the JHE produced during virus replication in tissue culture cells was secreted.

A reduction in feeding was observed only when the first instar *Trichoplusia ni* were infected with the JHE-recombinant virus as compared with the wild-type virus. No differences in feeding or weight gain were observed following infections of later instar larvae.

Bonning et al.[15] constructed an AcMNPV recombinant which expressed JHE under the control of a duplicated p10 promoter. Accordingly, the recombinant virus produced polyhedra. No significant differences in weight gain or LT_{50}s were detected following infections of 2nd instar *T. ni* larvae with the JHE-recombinant virus or the wild-type virus. A recombinant control virus containing bacterial lacZ gene rather than the JHE gene had a longer LT_{50}. This is consistent with the observation that larval death may be delayed with the production of beta-galactosidase.[16] The absence of pesticidal enhancement with the expression of JHE may have resulted from feedback regulatory systems which counteract the viral-induced JHE. Such feedback mechanisms may pose limitations to the use of insect hormones to enhance the pesticidal properties of baculoviruses.

Of particular note in the Bonning et al.[15] report is that the lacZ control and JHE recombinant viruses had 7.6- and 4.6-fold, respectively, higher LD_{50}s (dosage of polyhedra required to infect 50% of the larvae) for 2nd instar larvae than the wild-type virus. The expression of foreign gene(s) during replication is not expected to alter LD_{50}s. The increased LD_{50} may have arisen either by a reduction in the number of virions occluded within the polyhedra or by an alteration in the efficiency of infection. The former could result from interference by a foreign protein in the occlusion process. The later could occur if, during the construction of a recombinant virus, additional, undetected genomic alteration(s) occurred which influenced events such as attachment, penetration and/or uncoating. Either of these phenomena is noteworthy. Accordingly, the genetic alterations to recombinant baculoviruses must be assessed carefully in evaluating the potential environmental interactions and commercial potential of these viruses.

C.   ECDYSTEROID UDP-GLUCOSYL TRANSFERASE

Ecdysteroids are steroid hormones which are elevated both prior to molting to successive larval instars and to metamorphosis from the larval to pupal stage. The AcMNPV genome contains a gene coding for a protein with ecdysteroid UDP-glucosyl transferase (EGT) activity.[17,18] This enzyme inactivates ecdysteroids by transfer of a galactose molecule from UDP-galactose to C-22 of the ecdysteroid.[19]

With the expression of EGT during virus replication and the resulting inactivation of the ecdysteroid, the larvae are prevented from molting. Deletion of the EGT gene from the viral genome resulted in larval molting as well as the typical feeding cessation and wandering just prior to molts.

Fourth instar *Spodoptera frugiperda* larvae infected with EGT-minus virus exhibited slightly lower weight gains as compared to larvae infected with the wild-type (EGT-plus) virus. When the bacterial lacZ gene was removed from the EGT-minus recombinant virus, the deletion of EGT resulted in a significant reduction in food consumption and approximately 20% reduction in the LT_{50}.[20]

Additional studies were performed to determine if the expression of JHE by viruses lacking EGT expression could induce a premature pupal molt.[21] Larvae infected with EGT-minus virus that expressed the JHE exhibited weight gains and LT_{50}s similar to the EGT-minus recombinant viruses.

O'Reilly and Miller[20] compared the number of progeny polyhedra produced following infection of *S. frugiperda* larvae with an EGT-minus isolate of AcMNPV and wild-type AcMNPV. They showed that larvae infected with the EGT-minus virus exhibited lower weight gains and a 23% reduction in progeny polyhedra. It is not surprising that the number of progeny polyhedra would be altered with changes in the pesticidal properties of a virus. The number of polyhedra produced per larva is controlled by a multitude of factors including the level of inoculum, the method of inoculation and age of larvae.[22] Slight changes in any of these factors were shown to significantly change the mean yield of polyhedra per larva.

D.   SCORPION TOXIN

In the absence of significant alterations in pesticidal activity through viral regulated production of insect hormones, the use of insect specific neurotoxins were viewed as effective alternatives. The AaIT neurotoxin gene from the North African scorpion, *Andoctronus australis,* was cloned into AcMNPV[23,24] and the BmNPV.[25] The recombinant AcMNPV isolate expressing the scorpion toxin was constructed with polyhedrin genes, facilitating *per os* bioassays.

To affect secretion of the toxin from infected cells, Stewart et al.[23] fused the AaIT toxin with the AcMNPV gp67 secretory signal sequence. Using 2nd instar *T. ni* larvae, *per os* bioassays with the AaIT-Acmnpv exhibited a 25% reduction in LT_{50} as compared with wild-type virus infections. When 3rd instar larvae were infected with the AaIT-AcMNPV, there was a 50% reduction in feeding damage when compared with larvae infected with the wild-type AcMNPV.

McCutchen et al.[24] constructed an AaIT-AcMNPV which expressed the toxin fused to a silkworm bombyxin secretory signal. They bioassayed the virus with 2nd instar *Heliothis virescens* larvae. The AaIT-AcMNPV had a 30% lower LT_{50} as compared to the wild-type AcMNPV.

The AaIT gene was also inserted into the BmNPV genome through replacement of the polyhedrin gene. The AaIT-BmNPV was assayed by injection of nonoccluded virions into the hemocoel of 2nd instar *B. mori* larvae. Injection of 10^5 plaque-forming-units of recombinant virus resulted in cessation of feeding by 40 hours inoculation. The AaIT-BmNPV infections resulted in an approximately 40% faster time to death than infections with a control virus lacking both the AaIT and polyhedrin genes.

E.   STRAW ITCH MITE TOXIN

Tomalski and Miller[26] identified and cloned a gene from *Pyemotes tritici* which coded for the TxP-1 neurotoxin. Subsequently, they inserted the gene into a recombinant isolate of AcMNPV.[27] The TxP-1-AcMNPV lacked a polyhedrin gene and, therefore, 4×10^4 plaque-forming units were injected into 5th instar *T. ni* larvae. The time to death or paralysis with the TxP-1-AcMNPV infections was 40% faster than with wild-type AcMNPV infections.

F.   BT TOXIN

The *Bacillus thuringiensis* delta-endotoxin has long been recognized as an effective pesticide for the control of many agronomic insect pests. Merryweather et al.[28] and Martens et al.[29] inserted the HD-73 delta-endotoxin gene into the AcMNPV. The Bt-AcMNPV isolates were bioassayed in 2nd instar *Pieris brassicae* and *T. ni* larvae, respectively. In both studies the expression of the Bt toxin did not significantly alter the pesticidal properties of the virus.

It should be noted that the Bt toxins expressed in these studies were the Bt protoxin, a biologically inactive form. The protoxin normally undergoes cleavage in the insect gut giving rise to a truncated active toxin. The protoxin produced in the insect cells was shown to be biologically active when fed *per os* to susceptible larvae. Accordingly, cleavage of the protoxin probably did not occur in the infected cells, and, if cleavage did occur, there were no sites of action on the inner surface of the cell membrane.

G.   MAIZE MITOCHONDRIA FACTOR

The maize mitochondrial T-urf13 gene codes for the URF13 protein which is responsible for cytoplasmic male sterility and sensitivity to toxins produced by *Bipolaris maydis* race T and *Phyllosticta maydis*. The URF13 produced by a recombinant AcMNPV disrupted the integrity of the insect tissue culture cell membranes.[30] Bioassays were performed by injection of 3rd to 4th instar *T. ni* larvae with nonoccluded virus particles. The URF13-AcMNPV infections resulted in an approximate 40% reduction in time to death when compared to wild-type virus infections.

Based on the potential utility of foreign gene inserts constructed to date, the search for additional pesticidal genes is clearly a commercial priority. The insertion and expression of additional genes is performed very simply. The primary limitation in this area has been the availability of pesticidal genes. Clearly, recombinant viruses expressing insect neurotoxins have received the greatest interest. The reason for this has been that the neurotoxins have the potential for high specific activity and fast speed of action. In addition, it is possible to produce these genetically enhanced pesticides either *in vivo* or *in vitro*. Recombinant viruses expressing proteins which interfere with basic metabolic processes, such as the URF13-AcMNPV, may result in significant reductions in virus replication, making commercial production problematic.

III. ENVIRONMENTAL CONCERNS

As with synthetic chemical pesticides, naturally occurring and recombinant viral pesticides will be required to satisfy the requirements of regulatory agencies prior to registration. The potential hazards associated with chemical and viral pesticides differ mainly in the fact that viral pesticides replicate and, therefore, environmental contamination can increase in concentration and area.

There are several types of environmental concerns with respect to the commercialization of recombinant pesticides. The first is the potential interactions with humans and other vertebrates. With genetically enhanced viral pesticides, the direct interaction of the virus with a vertebrate host as well as indirect interaction of foreign gene products must be evaluated. Second, there must be an evaluation of the potential interaction of the pesticide with beneficial and other nontarget insect species. Unfortunately, the conclusions of most host range studies with wild-type baculoviruses have been based on exposure followed by observations of signs and symptoms.[31] Inapparent infections, i.e., virus replication in the absence of overt deleterious effects, were not investigated. There is a concern that the expression of foreign pesticidal genes with recombinant pesticides could change inapparent infections to lethal infections.

There are ecological concerns that the recombinant organisms may displace naturally occurring organisms from their niches in the environment. Environmental displacement may result in significant ecological perturbations. Despite all of the laboratory data which can be used to evaluate a recombinant product prior to release into the environment, there remains the possibility that any recombinant organism may be discovered to possess unwanted environmental/health properties. With recombinant viruses or other microorganisms, environmental mitigation could be problematic. Finally, the genetic stability of recombinant organisms must be evaluated. The primary concern is that a foreign gene inserted into virus "A" would be transferred to virus "B" with unanticipated and unwanted results.

These issues are substantive in nature. It must be realized, however, that zero environmental/health risk will probably never be achieved. As with synthetic chemical pesticides, there are risks. The goal of developing recombinant baculovirus pesticides should be to achieve risk factors that are lower than those associated with chemical pesticides. To achieve this goal, it is necessary to take a proactive approach rather than the reactive approach which has been previously taken with earlier pesticide evaluations. Although this requires higher developmental costs, the benefits of recombinant pesticides may clearly outweigh the potential direct and indirect costs associated with these alternative pest control strategies.

IV. FIELD TESTING OF RECOMBINANT BACULOVIRUSES

As part of the proactive approach to the development of genetically enhanced viral pesticides, two types of recombinant field testing programs have been conducted to date. The first has been designed to test the commercial potential of recombinant viruses under contained conditions. The second type has been the evaluation of genetic engineering strategies designed to limit the potential environmental concerns.

A.   MARKED VIRUS RELEASE

The first field testing of genetically engineered isolates of the AcMNPV was conducted in 1986 by scientists at the Natural Environment Research Council's (NERC) Institute of Virology in Oxford, U.K.[32] The release site was constructed to provide a high degree of physical containment and all residual virus was inactivated by formalin treatment of the soil at the conclusion of the experiment. The altered virus had a noncoding, 80-bp insert downstream of the polyhedrin gene coding region. The studies showed that the marked and parental isolates had identical host ranges, genetic stability during replication in cell cultures and stability in soil.

B.   POLYHEDRIN-MINUS VIRUS

In 1987 the NERC Institute of Virology conducted a second field test with a polyhedrin-minus isolate of AcMNPV.[32] The polyhedrin gene and promoter were replaced with a 100-bp insert. Laboratory-infected insects were placed on plants in the field enclosure and died within one week post-infection. An important observation in this experiment was that no virus infectivity could be detected on foliar or soil samples 7 days after larval death. In the absence of polyhedrin protein, the virus particles did not become occluded and were unstable following larval death. Accordingly, the marked polyhedrin-minus virus was called "self-destructive".

The self-destructive nature of the polyhedrin-minus virus was very attractive from an environmental standpoint. After killing the insect pest, there would be no residual virus. However, the instability of the virus particles meant that they could not be delivered to the field as an active product.

C.   CO-OCCLUDED VIRUS RELEASE

It was found that this instability problem could be overcome by a process called co-occlusion. The polyhedrin-minus virus particles can be occluded in polyhedra and stabilized by co-infecting cells with both the wild-type and polyhedrin-minus virus.[33-36] When the wild-type virus produced polyhedrin protein, the polyhedrin protein crystallized around both the wild-type and polyhedrin-minus virus particles, thereby forming polyhedra containing both virus types. The co-occlusion process effectively stabilized polyhedrin-minus virus isolates so they could be delivered to the field in an active form.

Laboratory studies showed that, during passage of co-occluded forms of AcMNPV from one insect to another, the progeny polyhedra population contained less and less of the polyhedrin-minus virus particles.[36] The nonpersistent nature of a polyhedrin-minus isolate of AcMNPV was field tested in 1989 by scientists at the Boyce

Thompson Institute and Cornell Agricultural Experiment Station.[37] This was the first field release of recombinant virus in the U.S. and, because of the biological containment provided by the removal of the polyhedrin gene, no physical containment or decontamination procedures were required.

In order to follow the persistence of the polyhedrin-minus virus in the environment, the co-occlusion field release was monitored for 3 years. The test site was originally sprayed with polyhedra which contain 49% polyhedrin-minus virus particles. In the 2nd and 3rd years, the progeny polyhedra contained 9 and 6% polyhedrin-minus virus, respectively.[37] Therefore, as predicted from the laboratory studies, the polyhedrin-minus virus had a very low potential to persist in the virus population as it cycled from one insect to the next.

A second co-occluded virus field release was initiated in 1993 with the *Lymantria dispar* MNPV.[38] The polyhedrin gene of the recombinant virus was replaced with a bacterial lacZ gene. The production of beta-galactosidase is being used to monitor the recombinant virus both in time and space. The test is being conducted in an oak forest, and gypsy moth is the target pest. The test will be completed in 1996.

It was discovered during the AcMPNV co-occlusion virus study that the soil in the test site contained high concentrations of biologically active polyhedra. The biological activity of these polyhedra remained constant during the study. This created a concern that, under agronomic conditions, the virus concentration would continue to increase in the environment with unknown consequences.

D. PRE-OCCLUDED VIRUS RELEASE

In order to preclude any potential environmental problems associated with high levels of virus contamination, a suicide strategy was developed which resulted in zero environmental contamination. The strategy took advantage of previously unrecognized properties of polyhedrin-minus viruses.

Late in the replication cycle of occluded baculoviruses, the viral nucleocapsids become membrane bound within the nucleus. The polyhedrin protein crystallizes around these membrane-bound particles, forming polyhedra. In the absence of a polyhedrin gene and polyhedra formation, the membrane-bound particles accumulate in the nucleus. These particles which were destined to become occluded are called pre-occluded virions (POV). The pre-occluded virions are highly infectious *per os* with susceptible host larvae.[16] The POV can be produced *in vitro* and are stable at 4°C.[16] They can also be produced *in vivo* if the larvae are freeze-dried prior to death.[39] The rehydrated tissues contain high concentrations of POV that are infectious *per os*.

It is possible, therefore, to produce inocula *in vivo* or *in vitro* which can be used to infect larvae *per os*. Field testing has been conducted for 2 years with the POV form of AcMNPV in cabbage plots infested with *T. ni* larvae. No foreign genes were inserted in the virus, only the polyhedrin gene was removed. As observed by Bishop et al.,[32] following death of the infected larvae in the laboratory or under field conditions, all infectivity of the progeny virus is lost within 7 days post-death. In the absence of stabilizing conditions, the nonoccluded POV are inactivated in the decaying larval tissues, resulting in zero environmental contamination or persistence.

The POV field studies were used to evaluate the infectivity of POV as compared to polyhedra. The tests compared the infectivity of POV and polyhedra produced in equivalent dry weights of infected larval tissue. In the first test, the POV preparations infected slightly more *T. ni* larvae than the polyhedra. In two additional tests, the POV preparations were slightly less active than the polyhedra samples.[39]

E.   AaIT Virus Release

In 1993 the NERC Institute of Virology conducted the first field study of a genetically improved viral pesticide.[40] The test involved an isolate of AcMNPV which expressed the AaIT scorpion toxin. Because of the expression of the scorpion toxin and the lack of a biological containment strategy, the field testing was performed under strict conditions of physical containment, and the chamber areas were formalin treated at the end of the experiment.

The field data indicated that the expression of AaIT toxin reduced the mean time to death of *T. ni* larvae by approximately 12% as compared to a 25% reduction under laboratory conditions.[23] At the lowest dosage, infection with the AaIT-expressing virus resulted in a 29% reduction in feeding damage to cabbage plants as compared to the larvae infected with the wild-type virus. Using 10- or 100-fold lower dosages, there was a smaller reduction in feeding damage as compared to the wild-type virus. This was the first field study to show that genetically enhanced viral pesticides can exhibit marked improvements over the pesticidal properties of naturally occurring viral pesticides.

V.   SUMMARY

To date, very few foreign genes have been evaluated for their potential to enhance the pesticidal properties of baculoviruses. The results clearly indicate that future research in this area will lead to the development of recombinant pesticides with properties equivalent to many of the synthetic chemical pesticides.

Although efficacy is a major concern in the pesticide industry, the cost/benefit ratio is the major factor which regulates the commercial development of pesticides. Accordingly, there are a multitude of economic factors which will determine the utility of viral pesticides.

Among these factors is the current state of the synthetic chemical pesticide industry and governmental regulatory agencies. The potential health/environmental concerns associated with chemical pesticides have led to the development of comprehensive testing procedures and safety requirements for the registration of pesticides. Because of this, it may cost from $30 to $60 million to develop and commercialize a single chemical pesticide. This cost is at least partially responsible for the significant reduction in the number of newly registered pesticides over the last 20 years.

Not only are fewer new pesticides coming to market, but there has also been a reduction in registration of products for specific crops. The pesticides registered under older and less stringent testing requirements are now being required to meet current standards. The pesticide industry has been forced to evaluate the costs of the additional testing vs. the potential profits. This has resulted in many pesticides not being registered for particular crops, mostly minor crops. This has resulted in a need for alternative pesticide strategies and opens the markets for biological control agents.

The costs to develop and register a wild-type baculovirus pesticide is estimated at less than a million dollars. In the U.S., six wild-type baculovirus pesticides have been registered: the *Helicoverpa zea* MNPV, *Orgyia pseudotsugata* MNPV, *Lymantria dispar* MNPV, *Neodiprion sertifer* SNPV, *Spodoptera exigua* MNPV and AcMNPV. The additional cost to register a recombinant viral pesticide will depend primarily upon the health/environmental attributes associated with the properties of the foreign pesticidal protein.

However, in recognition of the need for alternative pest control strategies and the potential benefits of biological control agents, regulatory agencies have adopted policies which encourage their commercial development. For instance, in 1994, the U.S. Environmental Protection Agency established a registration group charged with expediting the evaluation and registration of "safer pesticides." Consequently, the development of genetically enhanced baculovirus pesticides is being influenced by a multitude of scientific and economic factors. Clearly, recombinant technologies can significantly improve the pesticidal properties of these viruses. In addition, an economic climate appears to be developing which provides biological control agents with competitive cost/benefit ratios.

It should be recognized that recombinant viral pesticides will be required only for certain markets, such as markets requiring fast-acting pesticides because cosmetic damage cannot be tolerated. Clearly, these types of agronomic markets will require genetically enhanced baculovirus pesticides. However, there are other market niches in which wild-type viral pesticides may be equally appropriate, such as the control of many forest pests.

In conclusion, the need for alternative pesticide strategies in agriculture and forestry has created opportunities for the development and commercialization of wild-type and genetically enhanced viral pesticides. The current economic trends and evolving scientific advances make these biological control agents beneficial from the health/environmental perspective as well as an economic standpoint.

VI.   ACKNOWLEDGMENTS

Preparation of this manuscript was partially supported by Grant BCS-9111091 from the National Science Foundation, Grant 94-33120-0307 from the U.S. Dept. of Agriculture and Grant CR-822883-01 from the U.S. Environmental Protection Agency.

REFERENCES

1. Bilimoria, S. L., Taxonomy and identification of baculoviruses, in *The Biology of Baculoviruses*, Vol. 1, Granados, R. R. and Federici, B. A., Eds., CRC Press, Boca Raton, FL, 1986, chap. 2.

2. Huber, J., Use of bacuolviruses in pest management programs, in *The Biology of Baculoviruses*, Vol. 2, Granados, R. R. and Federici, B. A., Eds., CRC Press, Boca Raton, FL, 1986, chap. 7.
3. Moscardi, F. and Sosa-Gomex, D. R., A case study in biological control: soybean defoliating caterpillars in Brazil, in *International Crop Science I*, Buxton, D. R., Shibles, R., Forsberg, R. A., Blad, B. L., Asay, K. H. J., Paulsen, G. M., and Wilson, R. F., Eds., Crop Science Society of America, Madison, WI, 1993, chap. 17.
4. Mitsuhashi, J., Devlopment of insect cell culture media for biotechnology, in *Proc. Eighth Int. Conf. Invertebrate and Fish Tissue Culture*, Fraser, M. J., Ed., Tissue Culture Association, Columbia, MD, 1991, 83.
5. Carr, C. C. and Kolodny-Hirsch, D. M., Method and apparatus for mass producing insects entomopathogens and entomoparasites, U.S. Patent 5,178,094, 1993.
6. Hughes, P. R., High density rearing system for larvae, U.S. Patent 5,351,643, 1994.
7. Groner, A., Specificity and safety of baculoviruses, in *The Biology of Baculoviruses*, Vol. 1, Granados, R. R. and Federici, B. A., Eds., CRC Press, Boca Raton, FL, 1986, chap. 9.
8. Smith, G. E., Summers, M. D., and Fraser, M. J., Production of human beta interferon in insect cells infected with a baculovirus expression vector, *Mol. Cell. Biol.*, 3, 2156, 1983.
9. Pennock, G. D., Shoemaker, C., and Miller, L. K., Strong and regulated expression of *Escherichia coli* beta-galactosidase in insect cells with a baculovirus vector, *Mol. Cell. Bio.*, 4, 399, 1984.
10. Kataoka, H., Troetshler, R. G., Li, J. P., Kramer, S. J., Carley, R. L., and Schooley, D. A., Isolation and identification of a diuretic hormone from the tobacco hornworm, *Manduca sexta*, *Proc. Natl. Acad. Sci. U.S.A.*, 86, 2976, 1989.
11. Maeda, S., Increased insecticidal effect by a recombinant baculovirus carrying a synthetic hormone gene, *Biochem. Biophys. Res. Commun.*, 165, 1177, 1989.
12. Snyder, M., Hunkapiller, M., Yuen, D., Silvert, D., Fristrom, J., and Davidson, N., Cuticle protein genes of *Drosophila*: structure, organization and evolution of four clustered genes, *Cell*, 29, 1027, 1982.
13. Sparks, T. C. and Hammock, B. D., Comparative inhibition of the juvenile hormone esterases from *Trichoplusia ni, Tenebrio molitor*, and *Musca domestica, Pestic. Biochem. Physiol.*, 14, 290, 1980.
14. Hammock, B. D., Bonning, B., Possee, R. D., Hanzlik, T. N., and Maeda, S., Expression and effects of the juvenile hormone esterase in a baculovirus vector, *Nature, (London)* 344, 458, 1990.
15. Bonning, B. C., Hirst, M., Possee, R. D., and Hammock, B. D., Further development of a recombinant baculovirus insecticide expressing the enzyme juvenile homone esterase from *Heliothis virescens*, *Insect Biochem. Mol. Biol.*, 22, 453, 1992.
16. Wood, H. A., Trotter, K. M., Davis, T. R., and Hughes, P. R., *Per os* infectivity of preoccluded virions from polyhedrin-minus recombinant baculoviruses, *J. Invertebr. Pathol.*, 62, 64, 1993.
17. O'Reilly, D. R. and Miller, L. K., A baculovirus blocks insect molting by producing ecdysteroid UDP-glucosyl transferase, *Science*, 245, 1110, 1989.
18. O'Reilly, D. R. and Miller, L. K., Regulation of expression of a baculovirus ecdysteroid UDPglucosyltransferase gene, *J. Virol.*, 64, 1321, 1990.
19. O'Reilly, D. R., Brown, M. R., and Miller, L. K., Alteration of ecdysteroid metabolism due to baculovirus infection of the fall armyworm *Spodoptera frugiperda*: host ecdysteroids are conjugated with galactose, *Insect Biochem. Mol. Biol.*, 22, 313, 1992.

20. O'Reilly, D. R. and Miller, L. K., Improvement of a baculovirus pesticide by deletion of the EGT gene, *Bio/Technology,* 9, 1086, 1991.

21. Eldridge, R., O'Reilly, D. R., Hammock, B. D., and Miller, L. K., Insecticidal properties of genetically engineered baculoviruses expressing an insect juvenile hormone esterase gene, *Appl. Environ. Microbiol.*, 58, 1583, 1992.

22. Shapiro, M., Bell, R. A., and Owens, C. D., In vivo mass production of gypsy moth nucleopolyhedrosis virus, in *The Gypsy Moth: Research toward Integrated Pest Management*, Doane, C. C. and McManus, M. L., Eds., *U.S. Forest Serv. Tech. Bull. 1584*, 1981, p. 633.

23. Stewart, L. M. D., Hirst, M., Ferber, M. L., Merryweather, A. T., Cayley, P. J., and Possee, R. D., Construction of an improved baculovirus insecticide containing an insect-specific toxin gene, *Nature (London)*, 352, 85, 1991.

24. McCutchen, B. F., Chaudary, P. V., Crenshaw, R., Maddox, D., Kamita, S. G., Pelakar, N., Volrath, S., Fowler, E., Hammock, B. D., and Maeda, S., Development of a recombinant baculovirus expressing an insect-selective neurotoxin: Potential for pest control, *Bio/Technology,* 9, 848, 1991.

25. Maeda, S., Volrath, S. L., Hanzlik,T. N., Harper, S. A., Majima, K., Maddox, D. W., Hammock, B. D., and Fowler, E., Insecticidal effects of an insect-specific neurotoxin expressed by a recombinant baculovirus, *Virology*, 184, 777, 1991.

26. Tomalski, M. D. and Miller, L. K., Insect paralysis by baculovirus-mediated expression of a mite neurotoxin gene. *Nature (London),* 352, 82, 1991.

27. Tomalski, M. D. and Miller, L. K., Expression of a paralytic neurotoxin gene to improve insect baculoviruses as biopesticides, *Bio/Technology,* 10, 545, 1992.

28. Merryweather, A. T., Weyer, U., Harris, M. P. G., Hirst, M., Booth, T., and Possee, R. D., Construction of genetically engineered baculovirus insecticides containing the *Bacillus thuringiensis* ssp. *kurstaki* HD-73 delta endotoxin. *J. Gen. Virol.,* 71, 1535, 1990.

29. Martens, J. W. M., Honee, G., Zuidema, D., vanLent, J. W. M., Viss, B., and Vlak, J. M., Insecticidal activity of bacterial crystal protein expressed by a recombinant baculovirus in insect cells, *Appl. Environ. Microbiol.,* 56, 2764, 1990.

30. Korth, K. L. and Levings, C. S., III, Baculovirus expression of the maize mitochondrial protein URF13 confers insecticidal activity in cell cultures and larvae, *Proc. Natl. Acad. Sci. U.S.A.*, 90, 3399, 1993.

31. Wood, H. A. and Hughes, P. R., Biopesticides, *Science,* 261, 277, 1993.

32. Bishop, D. H. L., Entwistle, P. F., Cameron, I. R., Allen, C. J., and Possee, R. D., Field trials of genetically-engineered baculovirus insecticides, in *The Release of Genetically-Engineered Micro-Organisms*, Sussman, M., Collins, C. H., Skinner, F. A., and Stewart-Tull, D. E., Eds., Academic Press, New York, 1988, chap. 12.

33. Miller, D. W., Genetically engineered viral insecticides: practical considerations, in *Biotechnology for Crop Protection*, Hedin, P. A, Hollingworth, R. M., and Mann, J. J., Eds., American Chemical Society, Washington, D.C., 1988, chap. 31.

34. Shelton, A. M. and Wood, H. A., Microbial pesticides, *The World & I*, October 1989, 365.

35. Wood, H. A., Hughes, P. R., van Beek, N., and Hamblin, M., An ecologically acceptable strategy for the use of genetically engineered baculovirus pesticides, in *Insect Neurochemistry and Neurophysiology*, Borkovec, A. B. and Masler, E. P., Eds., Humana Press, Clifton, NJ, 1990, 285.

36. Hamblin, M., van Beek, N. A. M., Hughes, P. R., and Wood, H. A., Co-occlusion and persistence of a baculovirus mutant lacking the polyhedrin gene, *Appl. Environ. Microbiol.,* 56, 3057, 1990.

37. Wood, H. A., Hughes, P. R., and Shelton, A. M., Field studies of the co-occlusion strategy with a genetically altered isolate of the *Autographa californica* nuclear polyhedrosis virus, *Environ. Entomol.*, 23, 211, 1993.
38. Wood, H. A., unpublished data, 1994.
39. Wood, H. A. and Hughes, P. R., unpublished data, 1994.
40. Cory, J. S., Hirst, M. L., Williams, T., Hails, R. S., Goulson, D., Green, B. M., Carty, T. M., Possee, R. D., Cayley, P. J., and Bishop, D. H. L., Field trial of a genetically improved baculovirus insecticide, *Nature (London)*, 370, 138, 1994.

7 Molecular Biology of Bacteria for the Biological Control of Insects

Francis Rajamohan and Donald H. Dean

TABLE OF CONTENTS

I. INTRODUCTION

Insecticidal microorganisms were first reported in the 19th century in an unfavorable light when diseases of honey bees and silk moths were traced to microbial origin.[1] Among the biological pesticides, bacteria are the most potential and successful group of organisms identified so far for the effective control of insect pests and vectors of diseases that have rapidly become resistant to chemical pesticides.

Insecticidal bacteria may be broadly classified as true pathogens, opportunistic or facultative pathogens, and food poisoning organisms. Davidson[2] considers those microbes "which invade host tissue and overgrow the host" as the true pathogens and those "which kill the host by means of toxins produced outside the host" as food poisoning organisms. The best examples of bacterial insecticides that have been investigated for almost 85 years are the members of the genus *Bacillus*. Among these organisms, the most commonly used and intensively investigated insect pathogens are *Bacillus thuringiensis* and *Bacillus sphaericus*. The emphasis of this review is molecular biology; therefore, we shall necessarily limit our discussion to these two organisms.

II. *BACILLUS THURINGIENSIS* (BT)

In the early 20th century, Berliner[3] provided the first inkling that microbes could control insect pests when he isolated and named *Bacillus thuringiensis* (BT) from a diseased moth population in a German granary. *B. thuringiensis* is a Gram-positive, sporulating bacterium differing from *Bacillus cereus* only by the synthesis of several insecticidal crystal proteins (ICPs). BT is a complex species spanning more than 30 varieties of flagellar serotypes based on serological and biochemical tests. The ICPs are synthesized during the late growth phase of the bacteria and are accumulated in the cytoplasm as crystals or inclusion bodies. The crystal may account for up to 30% of the dry weight of the sporulated culture. The ICPs produced by BT are alpha-, beta-, and gamma-exotoxins and delta-endotoxins. Among these proteins, the delta-endotoxin and beta-exotoxin are used in agriculture. The delta-endotoxin is the most extensively studied toxin; its larvicidal specificity includes members of lepidopteran, dipteran, and coleopteran insects.

A.   LOCALIZATION OF ICP GENES

In the early 1980s, intensive research was directed toward the localization of the ICP genes among different BT subspecies. A number of early studies revealed direct correlation between the presence of plasmids and insecticidal crystal production. Plasmid curing, conjugation (mating), and gene cloning experiments provided further evidence that the ICP genes are very often located in larger (>30 MDa), low copy-number cryptic plasmids.[4-6] However, there are conflicting reports about the possible existence of ICP genes in the chromosome of some subspecies such as *kurstaki* HD1, *entomocidus, aizawai* 7.29, *dendrolimus,* and *wuhanensis.*[7-9] Carlton and Gonzalez[10] made a survey of the plasmids present in most of the BT strains. Their studies on 21 BT subspecies revealed that the number of plasmids varied from two to twelve between strains. Gene hybridization experiments provided additional evidence for the existence of multiple toxin-producing genes within a single bacterium. For example, the native plasmids of *B. thuringiensis* subspecies *aizawai* and *kurstaki* HD1 have as many as five separate ICP genes located at multiple sites.[11] In the case of *B. thuringiensis* subspecies *israelensis,* four dipteran-specific toxin genes and one cytolysin coding gene are located on a single 72 MDa plasmid.[8,9] The earlier studies have also shown that several lepidopteran-specific toxin genes (e.g., *B. thuringiensis* subspecies *kurstaki* HD73, *B. thuringiensis berliner* 1715) are flanked by two sets of insertion sequences and a transposable element.[12-14] This structural organization of these toxin genes might give an evolutionary advantage to the bacteria by enabling them to adapt and proliferate among several insect species.

Following the localization of toxin genes, many research groups began to clone ICP genes from several *B. thuringiensis* subspecies. Schnepf and Whiteley[15] reported the first cloning and expression of a *cryI* type gene from *B. t. kurstaki* HD1 "Dipel" plasmid preparation. They expressed the ICP gene in *Escherichia coli* and showed its activity to *Manduca sexta*. Following this report, protoxin genes were cloned and sequenced from numerous other BT species toxic to the members of the insect orders Lepidoptera, Diptera, and Coleoptera. At the present time, over 48 ICP gene sequences are given in gene databases or the journal and patent literature.

B.   CLASSIFICATION OF ICP GENES

Held et al.[16] first used the abbreviation "*cry*" (for crystal) to represent the insecticidal crystal producing genes of BT strains. Based on the primary target insect specificity and sequence similarity, insecticidal toxins have been classified into five major classes.[17,18] Table 1 shows a summary of BT crystal protein genes and insect specificity. All the genes encoding 130–140 kDa protoxin active against lepidopteran larvae are grouped into *cryI* class, which is further divided into several subclasses (A to G). Based on amino acid identity (>80%), the *cryIA* gene subclass has been further divided in to *IAa, IAb*, and *IAc*. The *cryII* type genes produce 66 kDa protoxin which is active against either lepidoptera alone (*cryIIB*) or lepidopteran and dipteran larvae (*CryIIA*). The *cryIII* genes produce a 73 kDa protein active against coleopteran larvae. The *cryIV* type genes have been isolated from the subspecies *israelensis* and produce 135, 128, 74, and 72 kDa proteins active against dipteran larvae. Recently, another novel gene has been isolated from *B. thuringiensis* subsp. *thompsoni* and named as *cryV*. This gene produces an 80 kDa toxin which is active to both lepidopteran and coleopteran larvae.[18] Unfortunately, several unrelated new *cry* genes have also been called *cryV*.[51] The name *CryVI* has been given to a group of genes that exhibit activity against nematodes.[52] Prior to this report all *B. thuringiensis* proteins that have identified have been active only against insects. This observation raises the question of the true extent of the host range of BT.

There are numerous inconsistencies in the current nomenclature. CryI is used only for toxins active against lepidopterans, but a mutant of CryIAb has been reported to be active against mosquito larvae as well.[25] CryII is reserved for toxins of dual activity (lepidopteran and dipteran), yet CryIIB is only active against lepidopterans.[39] CryIV is reserved for dipteran active toxins, but recently it has been observed that *B. thuringiensis* var. *israelensis*, which contains only *cryIV* genes, is active against a lepidopteran. The inconsistencies in the current nomenclature have led to a number of sessions at major meetings on *B. thuringiensis* concerning revisions of the nomenclature.

TABLE 1

Classification of B. *thuringiensis* delta-Endotoxin Genes and Specificity

cry Gene	Bt Subspecies	Target Insect		Refs.
		Order	Species	
crylAa	HD1, sotto, entomocidus	L	Bombyx mori, Mandura sexta	19–21
crylAb	kurstaki, HD1, berliner NRD12, IC1, IPL7	L/D	M. sexta, Pieris brassicae Aedes aegypti	22–26
crylAc	HD73, BTS89A	L	Heliothis virescens, Tricoplusia ni	27, 28
crylB	HD2	L	P. brassicae	29
crylC	aizawai, entomocidus	L	Spodoptera littoralis	30, 31
crylCb	galleriae	L	Spodoptera exigua	32
crylD	HD68	L	S. exigua, M. sexta	33
crylE	kenyae	L	S. littoralis	34
crylF	EG6346	L	Ostrinia nubilalis, S. exigua	35
crylG	galleriae, DSIR517	L		36, 37
crylX	galleriae	L		38
crylIA	HD1, HD263	L & D	Lymantria dispar, A. aegypti	39, 40
crylIB	HD1	L	L. dispar. T. ni	39
crylIC	shanghai S1	L	M. sexta, L. dispar	41
crylIIA	tenebrionis, san diego EG2158	C	Leptinotarsa decemlineata Phaedon cochleariea	42–44
crylIIB	tolworthi	C	L. decemlineata	45
crylIIC		C	Diabrotica undecimpunctata	46
crylIID	Bt1109P	C		47
crylVA	israelensis	D	A. aegypti, Culex pipiens	48
crylVB	israelensis	D	A. aegypti	49
crylVC	israelensis	D	A. aegypti	22
crylVD	israelensis	D	A. aegypti, C. pipiens	50
cryV	JHCC4835	L/C	O. nubilalis, Diabrotica spp.	18

Note: L, Lepidoptera; D, Diptera; C, Coleoptera.

C.   ACTIVATION OF PROTOXIN AND STRUCTURE-FUNCTION RELATIONSHIP

Upon ingestion of the CryI delta-endotoxins by the susceptible lepidopteran larvae, the crystalline inclusions are dissolved in the reducing alkali (pH >10) conditions of the larval midgut to yield the protoxin. In the following proteolytic activation, approximately half of the protoxin molecule is removed from the C terminus, leaving the N terminal half as the functional toxic domain. Activation also appears to be accomplished by the removal of approximately 28–30 residues from the N terminus. The CryI and IV type protoxins (130–140 kDa) are enzymatically processed into 65–75 kDa stable toxin by the susceptible larvae gut enzymes, whereas the protoxins of CryII and III (70–75 kDa) gene type are processed to 60–65 kDa active toxin. Based on the primary amino acid sequence, the amino-terminal halves are principally hydrophobic, whereas carboxy-terminal halves are primarily hydrophilic. The carboxyl half is richer in lysine and cysteine than the amino half. This observation suggests that this half might play a role in oligomerization of toxin and crystal formation through disulfide bridges. In general, most of the lepidopteran-active toxins can be divided into a highly conserved N-terminal region (aa 28–282), a middle hypervariable region (aa 283–461), and a partially conserved C terminus (aa 462–610). Another striking feature of the Cry toxins is the presence of five blocks of conserved amino acids among most of the insecticidal crystal proteins except CryII and CryIVD. These highly conserved tracts are considered to be important to protein function or structure.

The structure of delta-endotoxin remained uncertain until the crystal structure of CryIIIA was deduced by Li et al.[53] These authors have proposed that the Cry proteins may adopt the same protein folding scheme, since the core of CryIIIA molecule that holds the domain interfaces is built on the five highly conserved sequence blocks. The structure of CryIIIA (67 kDa) toxin is composed of three distinct domains. Domain I, composed of the N terminus 290 residues, is a bundle of seven α-helices in which the central hydrophobic helix (helix 5) is surrounded by six amphiphilic helices. Domain II, from residues 291 to 500, contains three anti-parallel β sheets each ending with a loop structure. Also, domain II comprises most of the hypervariable region of the toxin. Domain III is a β sandwich containing the conserved C-terminus region of the Cry toxins from residues 501 to 644. The function of the toxin domains will be discussed under the mode of action of the toxins.

D.   SPECIFICITY

The specificity of a toxin is its ability to exhibit toxicity to limited insect species or other pests. From another perspective, we may speak of insect susceptibility to a particular class of toxins. From either perspective, we must address the interaction of the toxin and the host. According to the current gene nomenclature, the first level of classification, the roman numeral denotes host range; that is, the specificity of the CryI group of toxins is against lepidopterans, as that of the CryIII toxins is against coleopterans. Even among toxins of the same classification, there is a significant difference in the degree of toxicity. For example, CryIAa is more active (400 times) against *Bombyx mori* than CryIAc, and CryIAc is more active (10 times) against *Heliothis virescens* or *Tricoplusia ni* than CryIAa.[54,55] Determination of the mechanisms of this specificity is of great interest, because the ability to change the specificity of a toxin would have potential impact on construction of second generation, broad spectrum toxins. For this purpose, Ge et al.[54] set out to locate the specificity-determining region of the CryIA toxins. They constructed a series of structurally stable proteins by exchanging smaller segments of amino acids between CryIAa and IAc toxins. With this protein engineering technique, Ge et al.[54] transferred the *B. mori* activity from CryIAa to IAc by exchanging amino acid residues

332–450 of CryIAa. This shows that the *B. mori* specificity of CryIAa toxin lies within the residues 332–450. Later, Ge et al.[55] revealed that the amino acids 335–450 on CryIAc are associated with the activity against *T. ni* , whereas residues 335–615 on the same toxin are required to exchange full *H. virescens* specificity. Recently, the lepidopteran (*Lymantria dispar*) and dipteran (*Aedes aegypti*) specificity-determining regions of CryIIA toxin were localized in residues 341–412 and 278–412, respectively.[56] Interestingly, most of the specificity-determining region of CryI toxins lies in the hypervariable region which corresponds to the domain II of CryIII toxin structure. Lee et al.[57]established a direct correlation between the specificity determining region and receptor binding in the case of *B. mori*. However, the "specificity" of a toxin and its receptor binding properties are not always directly correlated. The exceptional cases are discussed under receptor binding.

E.   MODE OF ACTION

After solubilization and proteolytic activation of the inclusions by sensitive larval midgut proteases, an active protease-resistant toxin core of 55–70 kDa is formed. This activation can be accomplished *in vitro* by trypsin digestion of the protoxin. The activated toxin binds to specific receptor molecules located in the microvillar brush border membranes.[58,59] After binding, the toxin irreversibly inserts into the membrane and alters the electrochemical potential gradient across the midgut by generating pores or selective/nonselective channels.[60,61] This destroys the osmotic balance of the cell membrane and causes the cell to swell and lyse. The important aspects of mode of action are receptor binding and channel formation.

1.   Receptor Binding

Several studies have established the fact that the recognition and binding of the toxin to a specific gut membrane molecule are essential for toxicity. Histopathological studies indicate that the toxin-binding target molecules are located in the microvillar brush border membranes which line the midguts of susceptible insects.[62] *In vitro* binding studies have been performed with brush border membrane vesicles (BBMV) prepared from larval midguts. Using [125]I- labeled toxins, saturable binding to the membrane receptor(s) with high binding affinity in several insects has been studied.[57-59] Most often, more than one such high affinity binding site is correlated with toxicity.[63]Several studies suggest a direct correlation between toxicity and the number of binding sites rather than binding affinity. However, some studies have failed to demonstrate a direct correlation between larvicidal activity and either the affinity or the number of binding site concentrations. For example, CryIA toxins do bind saturably to the BBMV prepared from *Spodoptera* species even though the toxin does not kill the larvae.[63] In another case, an inverse correlation between toxicity and membrane binding of CryIAc toxin to *Lymantria dispar* has been shown.[64]

The summation of the specificity, binding, and structural studies indicates that domain II of the toxin is involved in binding to the receptor. Recent experiments in our laboratory demonstrate that the loops of domain II are contact points in receptor binding,[65,66] as predicted by Li et al.[53] An exception to this is the report by Wu and Aronson[67] that the mutation A92D (domain I) in the *crylAc* gene radically disrupts receptor binding, as assessed by heterologous competition, to *M. sexta* BBMV (but only slightly to *H. virescens* BBMV). Since domain I is predicted to be involved only in channel formation, we have reexamined this location by constructing single amino acid substitution mutation (A92 D and A92E) in CryIAa and CryIAc toxin. Surprisingly, we do not observe any effect on binding of these mutants to *M. sexta* BBMV, but in agreement with Wu and Aronson,[67] we observe a dramatic inhibition on toxicity.[68]

Despite the published data and theoretical predictions that domain II is the toxin receptor binding region, there are results concerning CryIAb and IAc suggesting that binding may also take place via residues in domain III. The amino acid sequence homology of CryIAb and IAc are virtually identical through residue 452 (domains I and II), yet these toxins show considerable difference in binding properties. The amino acid difference between these two proteins lies solely in domain III in the region between conserved block three and four. More recently, immunoblotting

TABLE 2

Molecular Masses of *B. thuringiensis* delta-Endotoxin Binding Membrane Proteins of Lepidopteran Insect Larvae

Insect	Btk HD1	Toxin Binding Membrane Protein(s) (kDa)				Refs.
		CrylAa	CrylAb	CrylAc	CrylC	
B. mori	220,150,130					69
				120		70
H. virescens		170	170	140, 120	40	71, 72
				155, 120, 103		71
				90, 63		71
H. zea		170	170	150, 140, 120		72
				155, 120, 103		72
				90, 63		71
L. dispar		180	180	120a		70
M. sexta			210			74
				120a		73
				120		75
S. litura	160					69
		150	150	125	40	72
S. littoralis				40, 120	40, 65	76
		160	125, 115	40		72
S. exigua		200, 180	200,180	130, 115	40	72
S. frugiperda				148		73

[a]   Identified as aminpeptidase-N.

procedures were developed to identify the receptor proteins.[69–76] To date, several toxin binding receptor molecules have been identified and purified (Table 2). The toxin binding (receptor) proteins are generally larger and sometimes glycoproteins. Recently, the CryIAc toxin binding protein for *M. sexta* has been identified as aminopeptidase-N.[73] The amino peptidase-N is a glycoprotein of 120 kDa size.[75]

2.   Ion Channel (Pore Forming) Function

The structure of the BT toxin has the very apparent feature that domain I is an alpha helical bundle consisting of 7 alpha helices. This domain is an obvious candidate for membrane insertion and involvement in ion channel function.[53] After toxin proteins bind to the receptor(s) on the surface of the midgut epithelial cells, they insert into the apical membrane[77] and form a pore or ion channel.[60,61] The distinction between the toxin forming a "pore" or an "ion channel" may at one level seem moot since one leading authority, Hille,[78] defines ion channels as macromolecular pores. Proponents of a pore-like function of BT toxins make a distinction from ion channels in that larger molecules, such as glucose, have been reported to pass into phospholipid vesicles.[79] The ion channel properties have been described as nonselective, in which Na^+ as well as K^+ have been reported as being passed through the toxin channel/pore.[79,80] Other authors have suggested that the toxins behave selectively as K^+ channels.[81,82] Schwartz et al.[83] have made the pivotal observation that the CryIC toxin in bilipid layers behaves as a selective K^+ ion channel in alkaline conditions, but is nonselective and may pass Na^+ and anions in neutral to acidic conditions. This observation may resolve the selectivity issue. Virtually all of the studies that report nonselectivity of the toxin have been done at neutral or acidic pH on phospholipid vesicles or tissue culture cells (which require a pH of 6.8) while the lepidopteran insect midgut is highly alkaline.[84]

To date, there is no direct evidence that domain I inserts into the membrane or is the structural basis of the channel or pore. To the contrary, recent evidence indicates that domain III also participates in ion channel function.[85] In this work, the alternating arginines (R) present in the conserved block of amino acid residues, **QRYRVRIRY**AS, found in domain III of virtually all BT toxins were changed into the highly similar amino acid, lysine. Chen et al.[85] observed that if the second or forth R residue is mutated to lysine there is a significant reduction in toxicity and an inhibition of short-circuit current as measured by voltage clamping of whole insect midgut. Further work in collaboration with other researchers has revealed that mutations in this region affect other properties of the toxin normally associated with ion channel function. Mutations in the arginines of this conserved block may dramatically block the K^+ uptake into *B. mori* BBMVs[86] and the same mutations also block the K^+ channel activity of the toxin in bilipid membranes as measured by patch clamp.[87] It appears that this conserved block of amino acids behaves as a regulator of ion channel function. The similarity between this conserved block of amino acids, which structurally is a beta strand with an arginine face, and the conserved S4 helix of all classic ion channels, which also has an arginine face, has not escaped us.

III. *BACILLUS SPHAERICUS* (BS)

Bacillus sphaericus is an aerobic, rod-shaped, endospore-forming saprophytic bacterium occurring in nature. Although commercial interest in *B. thuringiensis* subsp. *israelensis* is high on account of its broader spectrum of action (mosquito and black fly), its field stability has been low. The long-term persistence of *B. sphaericus* strains in the aquatic environment directed researchers to pay more attention to this species, both for basic research on genetics and mode of action as well as a potential candidate for development of commercial mosquito larvicides.

A.   GENERAL CLASSIFICATION

de Barjac et al.[88] have attempted to classify *B. sphaericus* (BS) strains, based on their mosquito larvicidal activity and serological cross-reactivity. Among the pathogenic strains, some are "highly lethal"(1593, 2362, 2297) and others are not so lethal (SSII-I) to mosquito larvae. Krych et al.[89] have classified BS based on DNA homologies, but could find no correlation between toxicity and DNA homology. Later, the DNA homology study of Yousten[90] provided five homology groups, with group II being divided into subgroups IIA and IIB. He also developed a bacteriophage typing protocol which allows grouping of the insecticidal strains into seven phage types. Most of the mosquito larvicidal strains are classified in groups 3 and 4. Thus, the isolates of *B. sphaericus* are genetically diverse. From the studies of Krych et al.,[89] it is evident that there is only a marginal difference in the G+C content of the DNA among the BS strains investigated. However, all the toxic strains had more than 79% DNA homology to *B. sphaericus* 1593, which is a well-characterized toxic strain.

B.   PLASMID PROFILE

Delta endotoxins of *B. thuringiensis* are generally associated with mega plasmids, which are reported to be present in low copy number.[91] On the other hand, the search for the presence of plasmids in BS has been incomplete and somewhat conflicting. The presence of small plasmids have been reported in some BS strains, especially 2297.[92] Davidson et al.[93] mentioned the existence of a large plasmid in *B. sphaericus* 1593 and 1881 but not in *B. sphaericus* 1691 or 2362. Abe et al.[94] examined several strains and found large plasmids in strains 1881 but not 1593 or 1691. In neither case was the size specified. Singer[95] identified a 75 MDa plasmid in both larvicidal and nonlarvicidal strains. However, no direct evidence has been reported to link the mosquito larvicidal activity with plasmids in *B. sphaericus*. Most of the BS mosquito larvicidal genes have been cloned from the total cellular DNA. Also, the lack of any detectable plasmids in the highly larvicidal strain 1691 suggests that the toxin genes are more likely chromosomal.

C.   CLONING OF MOSQUITO LARVICIDAL GENES

The first report of the cloning of a mosquito larvicidal gene from *B. sphaericus*

1593 was made by Ganesh et al.[96] The recombinant plasmid (pGspo3), containing a 3.7 kb insert, was toxic to *Culex pipiens* and *Anopheles stephensi* (LC_{50} of 1 ng and 2μg cellular protein per milliliter, respectively). Subsequently, it was noted that the clone had lost its larvicidal potential, although the DNA insert appeared to remain the same size. Moreover, the restriction map of the above gene in question does not match the sites of the other mosquito larvicidal genes of *B. sphaericus*. Hence, this gene is not considered to be a potential larvicidal gene. Several groups of workers have since cloned a 3.5 kb HindIII fragment from several highly toxic *B. sphaericus* strains (1593, 2362, 2297, and IAB59) coding for the 42 and 51 kDa peptides.[97,98] Hindley and Berry[99] cloned the gene coding only for the 42 kDa protein from *B. sphaericus* 1593 into *E. coli* and determined its sequence. The nucleotide sequence alignment of different clones showed that the nucleotides of the coding region are identical for the strains 1593, 2362, and 2317.3, whereas the strains 2297 and IAB59 differ from the previous strains by 25 and 7 nucleotides, respectively.

The gene organization of the 51 and 42 kDa toxins suggests that the genes coding for these two proteins are located in a single transcription unit in all the highly toxic strains and are separated by a spacer region of 174–176 nucleotides. The open reading frames of both the 51 and 42 kDa proteins are preceded by a ribosome binding site and termination codons at the end of each gene. More recently, another toxin gene coding for a 100.6 kDa protein has been cloned and sequenced from a low toxic strain *B. sphaericus* SSII-I.[100] This gene has no significant homology with the 51 and 42 kDa protein genes, but shares similar restriction enzyme sites with another toxin gene isolated from *B. sphaericus* 1593M.[101]

D. CELLULAR LOCATION OF BS TOXIN

Until recently, it was thought that *B. sphaericus* did not synthesize parasporal crystals like that of *B. thuringiensis*. However, Davidson and Myers[102] and Kalfon et al.[103] have identified some parasporal inclusions in some BS strains, notably the highly toxic strains. The polyhedral, mosquito larvicidal crystal is largely synthesized during sporulation except in strain SSII-I which synthesizes the toxin in the vegetative growth phase.[103] Unlike *B. thuringiensis* crystals, the inclusion bodies of BS are held in close proximity to the spores by a membrane envelope because of a deficiency in autolytic enzymes.[104]

E. PROPERTIES OF BINARY TOXIN

To date, two distinct toxins have been isolated from *B. sphaericus*, the 51 and 42 kDa, binary highly active toxin and a 100 kDa weakly active toxin. The antigenically distinct 51 and 42 kDa toxins are produced by the highly active strains (1593, 2362, 2297) and both are necessary for mosquito larvicidal activity.[105,106] Though the 42 kDa protein alone is toxic to cultured mosquito cell lines, it is not toxic to mosquito larvae up to a concentration of 200 μg protein per milliliter.[97] However, it regains its activity (LC_{50} of 21 ng protein per ml) when the purified 51 kDa protein is added. A more recent report indicated that only 42 kDa toxin is active to *Culex pipiens* when its gene was expressed and purified from *B. thuringiensis* SPL407.[107] This result is contradictory to an observation by Baumann and Baumann[108] that the 42 kDa toxin alone is not toxic to mosquito larvae when purified from *B. subtilis* DB104. A quantitative bioassay performed by Broadwell et al.[109] suggested an equal ratio of 51 and 42 kDa toxins for maximum larvicidal activity. The 100 kDa toxin, which is isolated from *B. sphaericus* SSII-I, is weakly toxic to mosquito larvae and is produced during vegetative growth. Its potency is about 1000 times less than that of the highly toxic strains and is more unstable.

F.   MODE OF ACTION AND STRUCTURE-FUNCTION RELATIONSHIP

Several N- and C-terminal deletion mutants have been constructed to identify the precise region of 51 and 42 kDa proteins required for toxicity. These results indicate that the amino acid residues between 32 to 395 and 17 to 353 of the 51 and 42 kDa proteins, respectively, are essential to retain complete mosquito larvicidal activity.[105,110-112] The two proteins do not share any significant sequence similarity except for four short conserved segments.

Though the complete mechanism of action of the toxin to the susceptible larvae is not understood, the established sequence of events is (1) ingestion of the inclusion bodies by the filter-feeding susceptible larvae; (2) solubilization of the crystal in the alkaline pH of the midgut; (3) activation of the 51 and 42 kDa toxin to 43 and 39 kDa toxins, respectively, by larval gut enzymes; (4) binding of the activated toxins to specific receptor molecules located at the gastric cecum (GC) and posterior midgut(PMG); and (5) internalization of the toxin followed by mitochondrial swelling and cell lysis. Since the 39 kDa toxin derived from 43 kDa toxin by trypsin or chymotrypsin or mosquito larval gut enzyme is toxic to the tissue culture-grown *C. cuinquefasciatus* cells, this protein is considered to be the toxic region. The role of the 51 kDa toxin remained unclear until recent studies were conducted using fluorescent-labeled toxins. These experiments have shown that the N-terminal region of the 51 kDa toxin is essential to direct the binding of the toxins to specific receptor molecules, and its C terminus is necessary to interact with the 42 kDa toxin. Very interestingly, both the toxins are essential for internalization. Neither 51 nor 42 kDa toxin can be internalized alone. Moreover, the specific binding and internalization process of the toxins is not found in nontoxic and toxin-resistant cell lines.[113-116] This suggests an inevitable, conformational change induced by the 51 and 42 kDa toxin is essential for internalization of the toxin and for toxicity. However, the exact mechanism of structural alterations and the amino acid residues involved in internalization need to be investigated.

The 100 kDa toxin of *B. sphaericus* SSII-I does not share homology with any other known insecticidal toxin.[117] However, the amino acid sequence of this toxin has an N terminal potential hydrophobic signal sequence and a transmembrane domain similar to that of SI subunit of pertussis toxin and the A subunit of cholera toxin. This strongly suggests that this 100 kDa toxin might exert its action by ADP-ribosylation like pertussis and diphtheria toxins. Recently Thanabalu et al.[117] demonstrated the ability of the 100 kDa toxin to ADP-ribosylate proteins found in *C. quinquefasciatus* cell extracts and also to carry out self ADP-ribosylation. The mechanisms of action of the 51, 42 kDa binary toxin and the 100 kDa toxin of *B. sphaericus* to susceptible mosquito larvae are different.

In conclusion, though new insights are emerging by the understanding of CryIIIA crystal structure, the actual mechanisms of action of BT and BS toxins are surprisingly complex and difficult to rationalize.

REFERENCES

1. Falcon, L. A., Use of bacteria for microbial control, in *Microbial Control of Insects and Mites*, Burges, H. D. and Hussey, N. W., Eds., Academic Press, New York, 1977, chap. 3.
2. Davidson, E. W., Microbiology, pathology and genetics of *B. sphaericus:* biological aspects which are important to field use, *Mosq. News*, 44, 147, 1984.
3. Berliner, E., Über de Schlaffsucht der Mehlmottenraupe, *Zeit. Gesamst*, 252, 3160, 1911.
4. González, J. M., Jr., Dulmage, H. T., and Carlton, B. C., Correlation between specific plasmids and endotoxin production in *Bacillus thuringiensis. Plasmid*, 5, 351, 1981.
5. Lereclus, D., Bourgouin, C., Lecadet, M. M., Klier, A., and Rapoport, G., Role, structure and molecular organization of the genes coding for the parasporal delta-endotoxins of *Bacillus thuringiensis*, in *Regulation of Procaryotic Development,* Smith, I., Slepecky, R. A., and Setlow, P., Eds., American Society for Microbiology, Washington, D. C., 1989, 255.
6. Whiteley, H. R. and Schnepf, H. E., The molecular biology of parasporal crystal body formation in *B. thuringiensis, Annu. Rev. Microbiol.*, 40, 549, 1986.
7. Held, G. A., Bulla, L. A., Jr., Ferrari, E., Hoch, J., Aronson, A. I., and Minnich, S. A., Cloning and localization of the lepidopteran protoxin gene of *Bacillus thuringiensis* subsp. *kurstaki, Proc. Natl. Acad. Sci. U.S.A.*, 79, 6065, 1982.
8. Aronson, A. I., Beckman, W., and Dunn, P., *Bacillus thuringiensis* and related insect pathogens, *Microbiol. Rev.*, 50, 1, 1986.
9. Lereclus, D., Arantes, O., Chauffaux, J., and Lecadet, M. M., Transformation and expression of a cloned delta-endotoxin gene in *Bacillus thuringiensis, FEMS Microbiol. Lett.*, 60, 211, 1989.
10. Carlton, B. C. and Gonzalez, J. M., Plasmids and delta-endotoxin production in different subspecies of *Bacillus thuringiensis*, in *Molecular Biology of Microbial Differentiation*, Hoch, J. A. and Setlow, P., Eds., American Society for Microbiology, Washington D.C., 1985, 246.
11. Sanchis, V., Lereclus, D., Menou, G., Chaufaux, J., and Lecadet, M. M., Multiplicity of delta-endotoxin genes with different insecticidal specificities in *Bacillus thuringiensis aizawai* 7.29, *Mol. Microbiol.*, 2, 393, 1988.
12. Kronstad, J. W. and Whiteley, H. R., Inverted repeat sequences flank the *Bacillus thuringiensis* crystal protein gene, *J. Bacteriol.*, 160, 95, 1984.
13. Lereclus, D., Ribier, J., Klier, A., Menon, G., and Lecadet, M. M., A transposon-like structure related to the δ-endotoxin gene of *Bacillus thuringiensis, EMBO J.*, 3, 2561, 1984.
14. Bourgouin, C., Delecluse, A., Ribier, J., Klier, A., and Rapoport, G. A., *Bacillus thuringiensis* subsp. *israelensis* gene encoding a 125-kilodalton larvicidal polypeptide is associated with inverted repeat sequences, *J. Bacteriol.*, 170, 3575, 1988.

15. Schnepf, H. E. and Whiteley, H. R., Cloning and expression of the *Bacillus thuringiensis* crystal protein gene in *Escherichia coli, Proc. Natl. Acad. Sci. U.S.A.,* 78, 2893, 1981.
16. Held, G. A., Bulla, L. A., Jr., Ferrari, E., Hoch, J., and Aronson, A. I., Cloning and localization of the lepidopteran protoxin gene of *Bacillus thuringiensis* subsp. *kurstaki, Proc. Natl. Acad. Sci. U.S.A.,* 79, 6065, 1982.
17. Höfte, H. and Whiteley, H. R., Insecticidal crystal proteins of *Bacillus thuringiensis, Microbiol. Rev.,* 53, 242, 1989.
18. Tailor, R., Tippett, J., Gibb, G., Pells, S., Pike, D., Jordan, L., and Ely, S., Identification and characterization of a noval *Bacillus thuringiensis* δ-endotoxin entomocidal to co-leopteran and lepidopteran larvae, *Mol. Microbiol.,* 6, 1211, 1992.
19. Schnepf, H. E., Wong, H. C., and Whiteley, H. R., The amino acid sequence of a crystal protein from *Bacillus thuringiensis* deduced from the DNA base sequence, *J. Biol. Chem.,* 260, 6264, 1985.
20. Shibano, Y., Yamagata, A., Nakamura, N., Iizuka, T., and Takanami, M., Nucleotide sequence coding for the insecticidal fragment of the *Bacillus thuringiensis* crystal protein gene, *Gene,* 34, 243, 1985.
21. Masson, L., Marcotte, P., Prefontaine, G., and Brousseau, R., Nucleotide sequence of a gene cloned from *Bacillus thuringiensis* subspecies *entomocidus* coding for an insecticidal protein toxic for *Bombyx mori, Nucleic Acids Res.,* 17, 446, 1989.
22. Thorn, L., Garduno, F., Thompson, T., Decker, D., Zounes, M., Wild, M., Walfield, A. M., and Pollock, T. J., Structural similarity between the Lepidoptera and Diptera-specific insecticidal endotoxin genes of *Bacillus thuringiensis* subsp. *kurstaki* and *israelensis, J. Bacteriol.,* 166, 801, 1986.
23. Geiser, M., Schweitzer, S., and Grimm, C., The hypervariable region in the genes coding for entomopathogenic crystal proteins of *Bacillus thuringiensis*: nucleotide sequence of the *kurhd1* gene of subsp. *kurstaki HD1, Gene,* 48, 109, 1986.
24. Höfte, H., de Greve, H., Seurinck, J., Jansens, S., Mahillon, J., Ampe, C., Vandekerckhove, J., Vanderbruggen, H., Van Montagu, M., Zabeau, M., and Vaeck, M., Structural and functional analysis of a cloned delta endotoxin of *Bacillus thuringiensis berliner* 1715, *Eur. J. Biochem.,* 161, 273, 1986.
25. Haider, M. Z. and Ellar, D. J., Functional mapping of an entomocidal δ-endotoxin. Single amino acid changes produced by site-directed mutagenesis influence toxicity and specificity of the protein, *J. Mol. Biol.,* 208, 183, 1989.
26. Oeda, K., Oshie, K., Shimizu, M., Nakamura, K., Yamamoto, H., Nakayama, I., and Ohkawa, H., Nucleotide sequence of the insecticidal protein gene of *Bacillus thuringiensis* strain *aizawai* IPL7 and its high level expression in *Escherichia coli, Gene,* 53, 113, 1987.
27. Adang, M., Staver, M. J., Rocheleau, T. A., Leighton, J., Baker, R. F., and Thompson, D. V., Characterized full-length and truncated plasmid clones of the crystal protein of *Bacillus thuringiensis* subsp. *kurstaki* HD-73 and their toxicity to *Manduca sexta, Gene,* 36, 289, 1985.
28. Dardenne, F., Seurinck, J., Lambert, B., and Peferoen, M., Nucleotide sequence and deduced amino acid sequence of a *cryIA(c)* gene variant from *Bacillus thuringiensis, Nucleic Acids Res.,* 18, 5546, 1990.
29. Brizzard, B. L. and Whiteley, H. R., Nucleotide sequence of an additional crystal protein gene cloned from *Bacillus thuringiensis* subsp. *thuringiensis, Nucleic Acids Res.,* 16, 2723, 1988.
30. Sanchis, V., Lereclus, D., Memou, G., Chaufaux, J., Guo, S., and Lecadet, M. M., Nucle-otide sequence and analysis of the N-terminal coding region of the *Spodoptera*-active δ-endotoxin gene of *Bacillus thuringiensis aizawai* 7.29, *Mol. Microbiol.,* 3, 229, 1989.

31. Honée, G., van der Salm, T., and Visser, B., Nucleotide sequence of crystal protein gene isolated from *B. thuringiensis* subspecies *entomocidus* 60.5 coding for a toxin highly active against *Spodoptera* species, *Nucleic Acids Res.*, 16, 6240, 1988.

32. Kalman, S., Kiehne, K. L., Libs, J. L., and Yamamoto, T., Cloning of a noval *cryIC*-type gene from a strain of *Bacillus thuringiensis* subsp. *galleriae*, *Appl. Environ. Microbiol.*, 59, 1131, 1993.

33. Höfte, H., Soetaert, P., Jansens, S., and Peferoen, M., Nucleotide sequence and deduced amino acid sequence of a new lepidoptera-specific crystal protein gene from *Bacillus thuringiensis*, *Nucleic Acids Res.*, 18, 5545, 1990.

34. Bossé, M., Masson, L., and Brousseau, R., Nucleotide sequence of a noval crystal protein gene from *Bacillus thuringiensis* subspecies *kenyae*, *Nucleic Acids Res.*, 18, 7443, 1990.

35. Chambers, J. A., Jelen, A., Gilbert, M. P., Jany, C. S., Johnson, T. B., and Gawron-Burke, C., Isolation and characterization of a noval insecticidal crystal protein gene from *Bacillus thuringiensis* subsp. *aizawai*, *J. Bacteriol.*, 173, 3966, 1991.

36. Smulevitch, S. V., Osterman, A. L., Shevelev, A. B., Kaluger, S. V., Karasin, A. I., Kadyrov, R. M., Zagnitko, O. P., Chestukhina, G. G., and Stepanov, V. M., Nucleotide sequence of a noval δ-endotoxin gene *cryIg* of *Bacillus thuringiensis* subsp. *galleriae*, *FEBS Lett.*, 293, 25, 1991.

37. Gleave, A. P., Hedges, R. J., and Broadwell, A. H., Identification of an insecticidal crystal protein from *Bacillus thuringiensis* DSIR 517 with significant sequence difference from previously published toxin, *J. Gen. Microbiol.*, 138, 55, 1992.

38. Shevelev, A. B., Suarinky, M. A., Karasin, A. I., Kogan, Y. N., Chegtukhina, G. G., and Stepanov, V. M., Primary structure of cryX**, the novel δ-endotoxin-related gene from *Bacillus thuringiensis* subsp. *galleriae*, *FEBS Lett.*, 336, 79, 1993.

39. Widner, W. R. and Whiteley, H. R., Two highly related insecticidal crystal proteins of *Bacillus thuringiensis* subsp. *kurstaki* possess different host range specificities, *J. Bacteriol.*, 171, 965, 1989.

40. Donovan, W. P., Dankocsik, C. C., Gilbert, M. P., Gawron-Burke, M. C., Groat, R. G., and Carlton, B. C., Amino acid sequence and entomocidal activity of the P2 crystal protein. An insect toxin from *Bacillus thuringiensis* var. *kurstaki*, *J. Biol. Chem.*, 263, 561, 1988.

41. Wu, D., Cao, X. L., Bay, Y. Y., and Aronson, A. I., Sequence of an operon containing a noval δ-endotoxin gene from *Bacillus thuringiensis*, *FEMS Microbiol. Lett.*, 81, 31, 1991.

42. Sekar, V., Thompson, D. V., Maroney, M. J., Bookland, R. G., and Adang, M. J., Molecular cloning and characterization of the insecticidal crystal protein gene of *Bacillus thuringiensis* var. *tenebrionis*, *Proc. Natl. Acad. Sci. U.S.A.*, 84, 7036, 1987.

43. Herrnstadt, C., Gilroy, T. E., Sobieski, D. D., Bennett, B. D., and Gaertner, F. H., Nucleotide sequence and deduced amino acid sequence of a coleopteran-active delta-endotoxin gene from *Bacillus thuringiensis* subsp. *sandiego*, *Gene*, 57, 37, 1987.

44. Donovan, W. P., González, J. M., Jr., Gilbert, M. P., and Dankocsik, C. C., Isolation and characterization of EG2158, a new strain of *Bacillus thuringiensis* toxic to coleoptera larvae, and nucleotide sequence of the toxin gene, *Mol. Gen. Genet.*, 214, 365, 1988.

45. Sick, A., Gaertner, F., and Wong, A., Nucleotide sequence of a coleopteran active toxin gene from a new isolate of *Bacillus thuringiensis* subsp. *tolworthi*. *Nucleic Acids Res.*, 18, 1305, 1990.

46. Lambert, B., Höfte, H., Jansens, S., Soetaert, P., and Peferoen, M., Novel *Bacillus thuringiensis* insecticidal crystal protein with a silent activity against coleopteran larvae, *Appl. Environ. Microbiol.*, 58, 2536, 1992.

47. Lambert, B., Theunis, W., Aguda, R., Van Avdenhove, K., Decock, C., Jansens, S., Sevrinck, J., and Peferoen, M., Nucleotide sequence of gene *cryIIID* encoding a novel coleopteran active crystal protein from strain BT109P of *Bacillus thuringiensis* subsp. *kurstaki, Gene,* 110, 131, 1992.

48. Ward, E. S. and Ellar, D. J., Nucleotide sequence of a *Bacillus thuringiensis* var. *israelensis* gene encoding a 130 kDa delta-endotoxin, *Nucleic Acids Res.,* 15, 7195, 1987.

49. Tungpradubkul, S., Settasatien, C., and Panyim, S., The complete nucleotide sequence of a 130 kDa mosquito-larvicidal delta-endotoxin gene of *Bacillus thuringiensis* var. *israelensis, Nucleic Acids Res.,* 16, 1637, 1988.

50. Donovan, W. P., Dankocsik, C. C., and Gilbert, M. P., Molecular characterization of a gene encoding a 72 kilodalton cytolytic polypeptide from *Bacillus thuringiensis* subsp. *israelensis, J. Bacteriol.,* 170, 4732, 1988.

51. Narva, K. E. et al., unpublished; sequences deposited in GenBank, 1992.

52. Feitelson, J. S., Payne, J., and Kim, L., *Bacillus thuringiensis:* insects and beyond, *Biol. Technology,* 10, 271, 1992.

53. Li, J., Carroll, J., and Ellar, D. J., Crystal structure of insecticidal δ-endotoxin from *Bacillus thuringiensis* at 2.5 Å resolution, *Nature,* 353, 815, 1991.

54. Ge, A. Z., Shivarova, N. I., and Dean, D. H., Location of the *Bombyx mori* specificity domain on a *Bacillus thuringiensis* δ-endotoxin protein, *Proc. Natl. Acad. Sci. U.S.A.,* 86, 4037, 1989.

55. Ge, A. Z., Rivers, D., Milne, R., and Dean, D. H., Functional domains of *Bacillus thuringiensis* insecticidal crystal proteins, *J. Biol. Chem.,* 266, 17954, 1991.

56. Liang, E. and Dean, D. H., Location of the lepidopteran specificity region in an insecticidal crystal protein CryIIA from *Bacillus thuringiensis., Mol. Microbiol.,* 13(4): 569–575.

57. Lee, M. K., Milne, R. E., Ge, A. Z., and Dean, D. H., Location of *Bombyx mori* Receptor binding on *Bacillus thuriengiensis* delta-endotoxin. *J. Biol. Chem.,* 267, 3115, 1992.

58. Hofmann, C., Vanderbruggen, H., Höfte, H., Van Rie, J., Jansens, S., and Van Mellaert, H., Specificity of *Bacillus thuringiensis* δ-endotoxin is correlated with the presence of high-affinity binding sites in the brush border membrane of target insect midguts. *Proc. Natl. Acad. Sci. U.S.A.,* 85, 7844, 1988.

59. Van Rie, J., Jansens, S., Höfte, H., Degheele, D., and Van Mellaert, H., Specificity of *Bacillus thuringiensis* δ–endotoxins. Importance of specific receptors on the brush border membrane of the mid-gut of target insects, *Eur. J. Biochem.,* 186, 239, 1989.

60. Wolfersberger, M.G., V-ATPase-energized epithelia and biological insect control, *J. Exp. Biol.* 172, 377, 1992.

61. Knowles, B.H. and Dow, J. A. T., The crystal δ-endotoxin of *Bacillus thuringiensis:* models for their mechanism of action on the insect gut, *Bioessays,* 15, 469, 1993.

62. Denolf, P., Jansens, S., Peferoen, M., Degheele, D., and Van Rie, J., Two different *Bacillus thuringiensis* delta-endotoxin receptor in the midgut brush border membrane of the European corn borer, *Ostrinia nubilalis* (Hübner) (Lepidoptera:Pyralidae), *Appl. Environ. Microbiol.,* 59, 1828, 1993.

63. Van Rie, J., Jansens, S., Höfte, H., Degheele, D., and Van Mellaert, H., Receptors on the brush border membrane of the insect midgut as determinants of the specificity of *B. thuringiensis* delta-endotoxins, *Appl. Environ. Microbiol.,* 56, 1378, 1990.

64. Wolfersberger, M. G., The toxicity of two *Bacillus thuringiensis* delta-endotoxins to gypsy moth larvae is inversely related to the affinity of binding sites on midgut brush border membranes for the toxins, *Experientia,* 46, 475, 1990.

65. Lu, H., Rajamohan, F., and Dean, D. H., Identification of amino acid residues of *Bacillus thuringiensis* δ-endotoxin CryIAa associated with membrane binding and toxicity to *Bombyx mori, J. Bacteriol.,* 176: 5554–5559.

66. Rajamohan, F. and Dean, D. H., unpublished data, 1994.
67. Wu, D. and Aronson, A. I., Localized mutagensis defines regions of the *Bacillus thuringiensis* δ-endotoxin involved in toxicity and specificity, *J. Biol. Chem.*, 267, 2311, 1992.
68. Chen, X. J. and Dean, D. H., unpublished data, 1994.
69. Indrasith, L. S. and Hori, H., Isolation and partial characterization of binding proteins for immobilized delta endotoxin from solubilized brush border membrane vesicles of the silkworm, *Bombyx mori*, and the common cutworm, *Spodoptera litura, Comp. Biochem. Physiol.*, 102B, 605, 1992.
70. Lee, M. K. and Dean, D. H., unpublished data, 1994.
71. Oddou, P., Hartmann, H., and Geiser, M., Identification and characterization of *Heliothis virescens* midgut membrane proteins binding *Bacillus thuringiensis* δ-endotoxins, *Eur. J. Biochem.*, 202, 673, 1991.
72. Oddou, P., Hartmann, H., Radecke, F., and Geiser, M., Immunologically unrelated *Heliothis* sp. and *Spodoptera* sp. midgut-proteins bind *Bacillus thuringiensis* CryIA(b) δ-endotoxins, *Eur. J. Biochem.*, 212, 145, 1993.
73. Sangadala, S., Walters, F. S., English, L. H., and Adang, M. J., A mixture of *Manduca sexta* aminopeptidase and phosphatase enhances *Bacillus thuringiensis* insecticidal CryIA(c) toxin binding and ^{86}Rb$^+$-K$^+$ efflux *in vitro, J. Biol. Chem.*, 269, 10088, 1994.
74. Vadlamudi, R. K., Tae H. J., and Bulla L. A., Jr., A specific binding protein from *Manduca sexta* for the insecticidal toxin of *Bacillus thuringiensis* subsp. *berliner, J. Biol. Chem.*, 268, 12334, 1993.
75. Knowles, B. H., Knight, P. J. K., and Ellar, D. J., N-Acetyl galactosamine is part of the receptor in insect gut epithelia that recognises an insecticidal protein from *Bacillus thuringiensis, Proc. Roy. Soc. Lond.*, B245, 31,1991.
76. Sanchis, V. and Ellar, D. J., Identification and partial purification of a *Bacillus thuringiensis* CryIC δ-endotoxin binding protein from *Spodoptera littoralis* gut membranes, *FEBS*, 316, 264, 1993.
77. Wolfersberger, M. G., Hofmann, C., and Luthy, P., Interaction of *Bacillus thuringiensis* delta-endotoxin with membrane vesicles isolated from lepidoperean larval midgut, *Zentralbl. Bakter. Mikrobiol. Hyg. Abtial.*, 15, 237, 1986.
78. Hille, B., *Ion Chanels of Excitable Membranes*, 2nd ed., Sinaver, Sunderland, MA, 1993.
79. Knowles, B. H. and Ellar, D. J., Colloid-osmotic lysis is a general feature of the mechanism of action of *Bacillus thuringiensis* δ-endotoxins with different insect specificity, *Biochim. Biophys. Acta*, 924, 509, 1987.
80. Schwartz, J. L., Garneau, L., Masson, L., and Brousseau, R., Early response of cultured lepidopteran cells to exposure to δ-endotoxin from *Bacillus thuringiensis*: involvement of calcium and anionic channels, *Biochim. Biophys. Acta*, 1065, 250, 1991.
81. Crawford, D. N. and Harvey, W. R., Barium and calcium block *Bacillus thuringiensis* subsp. *kurstaki* δ-endotoxin inhibition of potassium current across isolated midgut of larval *Manduca sexta, J. Exp. Biol.*, 137, 277, 1988.
82. Slatin, S. L., Abrams, C. K., and English, L., Delta-endotoxin form cation selective channels in planar lipid bilayer, *Biochem. Biophys. Res. Commun.*, 169, 765, 1990.
83. Schwartz, J. L., Garneau, L., Savaria, D., Masson, L., Brousseau, R., and Rousseau. E., Lepidopteran-specific crystal toxins from *Bacillus thuringiensis* form cation and anion-selective channels in planar lipid bilayer, *J. Membr. Biol.*, 132, 53, 1993.
84. Dow, J. A. T., Insect midgut function, in *Adv. Insect Physiol.*, 19, 187, 1986.
85. Chen, X. J., Lee, M. K., and Dean, D. H., Site-directed mutations in a highly conserved region of *Bacillus thuringiensis* δ-endotoxin affect inhibition of short circuit current across *Bombyx mori* midguts, *Proc. Natl. Acad. Sci. U.S.A.*, 90, 1993.

86. Wolfersberger, M. G., Chen, X. J., and Dean, D. H., unpublished data, 1994.

87. Schwartz, J. L., unpublished data, 1994.

88. de Barjac, H., Larget-Thiery, I., Dumanior, V. C., and Ripouteau, H., Serological classification of *B. sphaericus* strains on the basis of toxicity to mosquito larvae, *Appl. Microbiol. Biotechnol.*, 21, 85, 1985.

89. Krych, V., Johnson, J. L., and Yousten, A. A., Deoxyribonucleic acid homologies among strains of *Bacillus sphaericus. Int. J. Syst. Bacteriol.*, 30, 476, 1980.

90. Yousten, A. A., *Bacillus sphaericus*: microbiological factors related to its potential as a mosquito larvicide, *Adv. Biotechnol. Process.*, 3, 315, 1984.

91. Gonzalez, J., Brown, B., and Carlton, B., Transfer of *B. thuringiensis* plasmids coding for delta-endotoxin among strains of *B. thuringiensis* and *B. cereus, Proc. Natl. Acad. Sci. U.S.A.*, 79, 6951, 1982

92. Kalfon, A., Charles, J. F., Bourgouin, C., and de Barjac, H., Sporulation of *Bacillus sphaericus* 2297: an electron microscope study of crystal-like inclusion biogenesis and toxicity to mosquito larvae, *J. Gen. Microbiol.*, 130, 893, 1984.

93. Davidson, E. W., Spizizen, J., and Yousten, A. A., Recent advances in the genetics of *B. sphaericus, Proc. Int. Colloq. Invertebr. Pathol.*, Brighton, England, 1982, 14.

94. Abe, K., Faust, R. M., and Bulla, L. A., Jr., Plasmid deoxyribonucleic acid in strains of *B. sphaericus* and in *Bacillus moritai, J. Invertebr. Pathol.*, 41, 328, 1983.

95. Singer, S., Current status of the microbial larvicide *Bacillus sphaericus*, in *Biotechnology in Invertebrate Pathology and Cell Culture*, Maramorosch, K., Ed., Academic Press, San Diego, CA, 133, 1987.

96. Ganesh, S., Kamdar, H., Jayaraman, K., and Szulmajster, J., Cloning and expression in Es*cherichia coli* of a DNA fragment from *Bacillus sphaericus* coding for biocidal activity against mosquito larvae, *Mol. Gen. Genet.,*189, 181, 1983.

97. Baumann, P., Baumann, L., Bowditch, R. D., and Broadwell, A. H., Cloning of the gene for the larvicidal toxin of *Bacillus sphaericus* 2362: evidence for a family of related sequences, *J. Bacteriol.*, 169, 4061, 1987.

98. Baumann, L., Broadwell, A. H., and Baumann, P., Sequence analysis of the mosquitocidal toxin genes encoding 51.4 and 41.9 kilodalton proteins from *Bacillus sphaericus* 2362 and 2297, *J. Bacteriol.*, 170, 2045, 1988.

99. Hindley, J. and Berry, C., Identification, cloning and sequence analysis of the *Bacillus sphaericus* 1593 41.9 kDa larvicidal toxin gene, *Mol. Microbiol.*, 1, 187, 1987.

100. Thanabalu, T., Hindley, J., Jackson-Yap, J., and Berry, B., Cloning, sequencing, and expression of a gene encoding a 100 kilodalton mosquitocidal toxin from *Bacillus sphaericus* SSII-I, *J. Bacteriol.*, 173, 2776, 1991.

101. Souza, A. E., Rajan, V., and Jayaraman, K., Cloning and expression in *Escherichia coli* of two DNA fragments from *Bacillus sphaericus* encoding mosquito-larvicidal activity, *J. Biotechnol.*, 7, 71, 1988.

102. Davidson, E. W. and Myers, P., Parasporal inclusions in *B. sphaericus, FEMS. Microbiol. Lett.*, 10, 2, 1985.

103. Kalfon, A., Charles, J. F., Bourgouin, C., and de Barjac, H., Sporulation of *B. sphaericus* 2297: an electron microscopic study of crystal-like inclusion biogenesis and toxicity to mosquito larvae, *J. Gen. Microbiol.,*130, 893, 1984.

104. Holt, S. C., Gauthier, J. J., and Tipper, D. J., Ultrastructural studies of sporulation in *B. sphaericus, J. Bacteriol.*, 122, 1322, 1975.

105. Broadwell, A. H., Baumann, L., and Baumann, P., Larvicidal properties of the 42 and 51 kilodalton *Bacillus sphaericus* proteins expressed in different bacterial hosts: evidence for a binary toxin, *Curr. Microbiol.*, 21, 361, 1990.

106. Davidson, E. W., Oei, C., Meyer, M., Bieber, A. L., Hindley, J., and Berry, B., Interaction of the *Bacillus sphaericus* mosquito larvicidal proteins. *Can. J. Microbiol.*, 36, 870, 1990.

107. Nicolas, L., Nielsen-Leroux, C., Charles, J. F., and Delécluse, A., Respective roles of the 42 and 51 kDa components of the *Bacillus sphaericus* toxin overexpressed in *Bacillus thuringiensis, FEMS Microbiol. Lett.*, 106, 275, 1993.

108. Baumann, L. and Baumann, P., Effects of the components of the *Bacillus sphaericus* toxin on mosquito larvae and mosquito-derived tissue culture grown cells, *Curr. Microbiol.*, 23, 51, 1991.

109. Broadwell, A. H., Baumann, L., and Baumann, P., The 42- and 51-kilodalton mosquitocidal proteins of *Bacillus sphaericus* 2362: construction of recombinants with enhanced expression and *in vivo* studies of processing and toxicity, *J. Bacteriol.*, 172, 2217, 1990.

110. Clark, M. A. and Baumann, P., Deletion analysis of the 51-kilodalton protein of the *Bacillus sphaericus* 2362 binary mosquitocidal toxin: construction of derivatives eqivalent to the larva-processed toxin, *J. Bacteriol.*, 172, 6759, 1990.

111. Oei, C., Hindley, J., and Berry, C., An analysis of the genes encoding the 51.4 and 41.9 kDa toxins of *Bacillus sphaericus* 2297 by deletion mutagenesis: the construction of fusion proteins, *FEMS Microbiol. Lett.*, 72, 265, 1990.

112. Sebo, P., Bennardo, T., Torre, de la, and Szulmajster, J., Deletion of the minimal portion of the *Bacillus sphaericus* 1593M toxin required for the expression of larvicidal activity, *Eur. J. Biochem.*, 194, 161, 1990.

113. Oei, C., Hindley, J., and Berry, C., Binding of the purified *Bacillus sphaericus* binary toxin and its deletion derivatives to *Culex quinquefasciatus* gut: elucidation of functional binding domains, *J. Gen. Microbiol.*, 138, 1515, 1992.

114. Davidson, E. W., Binding of the *Bacillus sphaericus* (Eubacteriales: Bacillaceae) toxin to midgut cells of mosquito (Diptera: Culicidae) larvae: relationship to host range, *J. Med. Entomol.*, 25, 151, 1988.

115. Davidson, E. W., Variation in binding of *Bacillus sphaericus* toxin and wheat germ agglutinin to larval midgut cells of six species of mosquitoes, *J. Invertebr. Pathol.*, 53, 251, 1989.

116. Schroeder, J. M., Chamberlain, C., and Davidson, E. W., Resistance to the *Bacillus sphaericus* toxin in cultured mosquito cells, *In Vitro Cell. Dev. Biol.*, 25, 887, 1989.

117. Thanabalu, T., Berry, C., and Hindley, J., Cytotoxicity and ADP-ribosylating activity of the mosquitocidal toxin from *Bacillus sphaericus* SSII-I: possible roles of the 27 and 70 kilodalton peptides, *J. Bacteriol.*, 175, 2314, 1993.

8 Molecular Biology of Fungi for the Control of Insects

John M. Clarkson

TABLE OF CONTENTS

I. INTRODUCTION

Insect pathogenic fungi have considerable potential for the biological control of insect pests of plants. These fungi are widely distributed through the fungal kingdom (Eumycota), although the majority of those classified occur in the Deuteromycotina and Zygomycotina.[1] Many attempts have been made to exploit the Deuteromycotina fungi *Metarhizium anisopliae, Beauveria* spp., *Aschersonia aleyrodis, Verticillium lecanii, Nomuraea rileyi* and some Entomophthorales for insect control, and at present fungi are being used on a moderate scale in Brazil, China and Russia.[2] Small amounts of *V lecanii* and *B. bassiana* are sold in Europe for pest control and *M. flavoviride* is being field-tested in Africa for control of locusts and grasshoppers.[3] A number of factors currently limit the full exploitation of entomopathogenic fungi, one of which is our poor understanding of the biochemistry and molecular biology of fungal pathogenesis. Progress in this area will help the production of more efficient mycoinsecticides either by identifying those attributes which should be selected for in a development program, or by identifying genes which could be up-regulated or otherwise manipulated to enhance virulence. This chapter reviews the latest developments in our understanding of the biochemistry and cell and molecular biology of fungal pathogenesis in insects with particular focus on *Metarhizium anisopliae.*

II. INFECTION STRUCTURE FORMATION

Entomopathogenic fungi invade their hosts by direct penetration of the host exoskeleton or cuticle. This is composed of two layers, the outer epicuticle and the procuticle (Figure 1). The epicuticle is a very complex, thin structure which is devoid of chitin but contains phenol stabilized protein.[4] The procuticle constitutes the majority of the cuticle and comprises chitin fibrils embedded in a protein matrix, together with lipids and quinones. In many areas of cuticle the chitin is laid down helicoidally, giving rise to a laminate structure. As is common with many plant pathogenic fungi, conidia germinate on the host surface and differentiate an infection structure termed an appressorium (Figure 1). From here, an infection hypha penetrates down through the host cuticle and eventually emerges into the insect's haemocoel. The formation of the appressorium plays a pivotal role in establishing a pathogenic relationship with the host and therefore elucidating the environmental cues which trigger appressorium development is central to our understanding of the pathogenic process.

Several entomopathogenic fungi are able to produce appressoria *in vitro* on hard hydrophobic surfaces. Differentiation in *M. anisopliae* is stimulated by low levels of complex nitrogenous compounds and is subject to carbon-catabolite repression.[5] Appressorium formation *in vivo* is also influenced by surface topography, appressoria

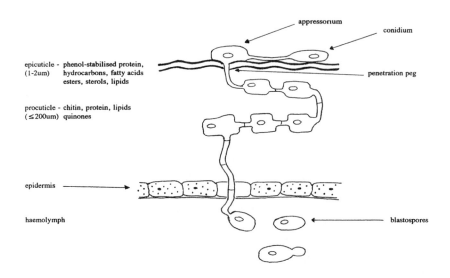

FIGURE 1.   Structure of insect cuticle and mode of penetration.

forming only after extensive growth over the microfolds of early 5th instar larvae of *Manduca sexta*, whereas germination on the relatively smooth surface of 5-day instar larvae allows appressorium formation close to the conidium. Similar results have been obtained using polystyrene cuticle replicates confirming a thigmotropic component to appressorium formation.[6]

Biochemical and pharmacological investigations have provided strong evidence for the involvement of cAMP in appressorium formation and several components of cAMP signal transduction pathways have been identified. Differentiation in *Metarhizium* is likely to be initiated following a contact-induced change in membrane potential, possibly due to the activation of a mechanosensitive ion channel. Consistent with this, gadolinium, a potent inhibitor of stretch-activated ion channels, inhibits appressorium formation by *M. anisopliae* at concentrations which have no effect on hyphal extension.[7] Mechanosensitive ion channels have been identified in the plant pathogen *Uromyces appendiculatus* and have been postulated to trigger differentiation of appressoria in this fungus.[8] In *Metarhizium* this transient membrane depolarization may result in conformational change and subsequent activation of GTP-binding proteins. Several families of G proteins have been identified in plasma membranes of *M. anisopliae,* including separate substrates for pertussis and cholera toxins, both of which were antigenically related to mammalian G proteins.[9] Plasma membranes also contain adenylate cyclase activity which is stimulated by the stable GTP analogue guanyl 5'-imidodiphosphate when *Metarhizium* membranes were reconstituted with a homogenate from a *crisp-l* mutant of *Neurospora crassa* lacking adenylate cyclase activity. This suggests that the adenylate cyclase activity of *Metarhizium* membranes is controlled by a labile regulatory G protein.[10] Intracellular levels of cAMP fluctuate but are higher in differentiating germlings, a significant rise in cAMP coinciding with the onset of appressorium formation.[11] Using a photoaffinity analogue of cAMP, one major high affinity cAMP-binding protein, the putative regulatory subunit of a cAMP-dependent protein kinase (PKA), has been identified in the plasma membrane.[10] Addition of the PKA inhibitor H8 selectively represses both the *in vitro* phosphorylation of a 27 kDa protein[10] and appressorium formation.[11]

The identity of the substrates for PKA in *Metarhizium* are unknown, but by analogy with higher eukaryotes, one might be CREB, the bZip class transcription factor which binds to cAMP-responsive elements (CREs) in the promoters of cAMP-inducible genes.[12] Interestingly the promoter of at least one gene which is highly induced during appressorium formation, *prl,* contains a CRE.[13]

A role for cAMP in appressorium development *in vitro* has also been demonstrated for the rice blast fungus *Magnaporthe grisea* where addition of cAMP to conidia germinating on a normally noninductive hydrophilic surface induced differentiation.[14] The PKA from this fungus has been cloned recently; subsequent transformation-mediated disruption of this gene specifically prevented appressorium formation on inductive surfaces, but did not affect hyphal growth, conidiation or sexual reproduction.[15]

Although the second messenger Ca^{2+} has been implicated in conidial germination,[16] its role in appressorium formation is less clear. Germlings continue to produce high levels of appressoria in a Ca^{2+}-free medium (containing EGTA), indicating that Ca^{2+} uptake is not required for differentiation.[11] However, the Ca^{2+} ionophore Br-A23187 plus EGTA prevents appressorium formation, presumably as a result of loss of intracellular Ca^{2+}. Simultaneous application of ionophore and exogenous Ca^{2+} substantially increases polar hyphal growth and completely inhibits appressorium formation, suggesting that a critical level of Ca^{2+} is required for differentiation.

III. CUTICLE PENETRATION

A range of extracellular enzymes capable of degrading the major components of insect cuticle is produced when *M. anisopliae* is grown *in vitro* with cuticle as the sole carbon and nitrogen source (Table 1). The precise role that individual enzymes play in cuticle penetration and subsequent stages of pathogenesis is poorly understood, although there is good biochemical evidence for the involvement of proteases. Although the complex structure of insect cuticle suggests that penetration would require the synergistic action of several different enzymes, much of the attention has focused on the cuticle-active endoprotease PR1 as a key factor in the process. Following the development of genetic transformation techniques for *Metarhizium*,[17] analysis of the contribution of individual gene products should be possible by targeted gene disruption .

The production of cuticle-degrading enzymes by *M. anisopliae* during infection structure formation has been investigated both *in vitro* and *in vivo*.[18,19] Because of their transparency and ready supply, insect wings provide an ideal form of cuticle for combined biochemical and histochemical analysis. Among the first enzymes produced on wings are endoproteases (PR1 and PR2) and aminopeptidase, coincident with the formation of appressoria. N-Acetylglucosaminidase is produced at a slow rate compared with proteolytic enzymes, and chitinase and lipase activities were not detected. *In situ* histochemical localization has confirmed that high levels of proteolytic enzymes are secreted by appressoria.[19]

The major protein synthesized during appressorium development *in vitro* is PR1, as evidenced by Western blotting and *in vivo* labelling with [^{35}S] methionine.[18] The addition of either glucose or alanine to conidia germinating *in vitro* repressed both appressorium formation and PR1, suggesting coordinate regulation by catabolite repression. Nutrient starvation is therefore likely to be a key environmental signal for the switch from a saprophytic to a pathogenic mode of growth, possibly following depletion of nutrients on the insect surface. Interestingly, isolates of *M. anisopliae* have been identified in which the differentiation of appressoria is not glucose repressed.[20] Some of these "catabolite derepressed" strains were originally isolated from hemipteran insects (plant sap-suckers) where nutrients on the cuticle are likely to be supplemented with insect secretions rich in sugars, suggesting that the regulation of key pathogenicity factors may be related to host specificity.

PR1 is not required for appressorium formation,[19] and the massive synthesis of PR1 during appressorium formation suggests developmental regulation possibly via the cAMP/PKA signalling pathway as discussed above.

TABLE 1

Cuticle-Degrading Enzymes of *M. anisopliae*

Enzyme	Specificity	Regulation	Refs.
PR1 subtilisin subclass serine endoprotease	Chymoelastase	C,N-repression, specific protein induction	27, 28, 71
PR2 serine endoprotease	Trypsin-like	C,N-repression, non-specific protein induction	26, 71
PR4 cystein endoprotease	Trypsin-like	?	72
Metallo aminopeptidase	Post alanyl	N-repression, weak C-repression, protein induction?	73
Serine dipeptidylpeptidase	Post prolyl	N-repression, weak C-repression, protein induction?	73
Chitinase	>NAG$_3$	C-repression NAG induction	35, 74, 75
N-Acetylglucosaminidase	NAG$_{2-4}$	Constitutive	35

Developmental regulation of extracellular depolymerases has also been observed during the formation of infection structures by the biotrophic plant pathogen *Uromyces viciae-fabae*, where cell wall-degrading enzymes are produced in a strictly differentiation-specific manner. However, as might be expected for a pathogen which lacks the ability for saprophytic growth, enzyme production is neither substrate inducible nor catabolite repressible.[21]

The involvement of PR1 in cuticle penetration is supported by ultrastructural localization using immunogold-labelling.[22] PR1 is secreted by appressoria and penetrating hyphae within the cuticle. Penetration of the epicuticle apparently occurs via enzymic degradation, as a high density of gold particles is present at the point of penetration which shows no signs of physical distortion. In contrast, penetration of the procuticle involved both mechanical and enzymic processes, as evidenced by the physical separation of lamellae by penetrating hyphae and secretion of PR1.

The question of the role of PR1 in pathogenicity has been tackled more directly using a specific inhibitor of PR1.[23] Simultaneous application of inhibitor and conidia to the insect surface causes a significant delay in mortality which is also associated with reduced disease symptoms and enhanced growth rate in surviving insects. The central role of endoproteases in the infection process is also reflected in their widespread distribution among entomopathogenic fungi. All isolates of *M. anisopliae* tested to date, as well as *Beauveria bassiana*, *Verticillium lecanii* and *Aschersonia aleyrodis* produce PR1 and PR2-like endoproteases.[24,25]

The regulation of *M. anisopliae* proteases in nondifferentiating broth cultures has been studied in detail.[26–28] Both PR1 and PR2 of isolate ME1 are subject to carbon catabolite and nitrogen metabolite repression. However, it has been shown recently that a unique form of regulation of PR1 occurs by specific induction by insect cuticle. Using ergosterol content as a measure of fungal growth on insoluble cuticle, it has been shown that the ca. 10-fold elevated levels of PR1 produced within 24 h of contact with cuticle resulted from induction and not just from an increase in fungal biomass. Growth but not PR1 induction occurs with elastin, BSA or gelatin. Deproteinated cuticle failed to induce PR1, whereas peptides released from cuticle by PR1 or PR2 induced PR1. This is the first example of specific induction of a microbial protease; the unique regulation of PR1 and its efficacy against insect cuticle presumably reflects adaptation to insect parasitism.

The *prl* cDNA has been cloned recently revealing that PR1 is synthesized as a large precursor containing a signal peptide, a propeptide and a mature 28.6 kDa protein.[29] The predicted amino acid sequence shows extensive similarity with the subtilisin subclass of serine proteases, the predominant class of microbial extracellular proteases. Southern blot analysis has demonstrated that genes with significant homologies to *prl* are present in the entomopathogens *A. flavus* and *Verticillium lecanii*. A gene similar to *prl* encoding the subtilisin-like serine protease PRB1 has been cloned recently from the mycoparasitic fungus *Trichoderma harziamum*. Intriguingly, this protease, which has the same identified substrate specificity as PR1 against synthetic peptides, is also induced in a host-related manner, in this case by fungal cell walls or chitin.[30]

The genomic *prl* promoter contains a number of putative binding sites for regulatory proteins homologous to CREA (C-catabolite repressor) and AREA (N-metabolite regulator) of *A. nidulans* and CREB, the eukaryote cAMP-responsive element binding protein. Using PCR primers based on the DNA-binding domain of CREA, a fragment has been amplified from *M. anisopliae* which shows 91% identity to CREA at the amino acid level, suggesting the presence of a *creA* homologue in this fungus.[13] Gel-retardation experiments with the *prl* promoter and a CREA-GST fusion protein have demonstrated that two of the three putative CREA binding sites are functional *in vitro*.

Chitin microfibrils constitute an important structural component of insect cuticle and therefore represent a potential barrier to penetration by entomopathogenic fungi. Consistent with this, it has been shown that dual application of a specific inhibitor of chitin synthesis and *M. anisopliae* have a synergistic effect.[31] Subsequent ultrastructural studies have demonstrated enhanced fungal penetration through inhibitor-treated cuticle.[32] However, in other studies, chitinase was not detected *in vivo* during the early stages of penetration[19] and was only detected substantially later than proteolytic enzymes *in vitro* on cuticle.[33] There is evidence that chitin degradation requires the prior action of proteases,[34] presumably due to the protein matrix surrounding chitin microfibrils in the cuticle. Resolving the contribution of chitinases to penetration is complicated by the multiplicity of extracellular isozymes produced by *M. anisopliae* and other entomopathogenic fungi.[35]

IV. TOXIN PRODUCTION

Entomopathogenic fungi produce a number of low molecular weight insecticidal toxins *in vitro*.[37] Destruxins (DTX) are a family of closely related cyclic depsipeptides produced by *M. anisopliae* and comprising five amino acids cyclised by the incorporation of a hydroxy acid; 17 variants have been described, most of which differ in the number of methylated amino acids and/or the type of hydroxy acid.[38,39] DTX are insecticidal by injection and toxicity is most acute among lepidopteran larvae and adult Diptera.[40] The toxins have shown diverse effects on various insect tissues and their functions. DTX depolarizes lepidopteran muscle membrane by directly or indirectly activating calcium channels.[41,42] This leads to tetanic, then flaccid, paralysis, which is reversible at low doses, implicating detoxification processes. Indeed, linearized and hydrated metabolites of DTX E have been found in the blood of *Galleria mellonella*[43] following injection of the native compound into the haemocoel. Mechanisms of detoxification may contribute to variation in species susceptibility to DTX.[40] A number of functions of insect haemocytes are inhibited *in vitro* by DTX,[44] as is the secretion of fluid by desert locust Malpighian tubules[45] and the production of ecdysterone by the prothoracic glands of larvae of the tobacco hornworm *Manduca sexta*.[46] In contrast to several other cyclic peptide antibiotics such as beauvericin, DTX has no ionophoric nor antimicrobial activity.[37] A number of the physiological actions of DTX on insect cells, most notably the effects on lepidopteran muscle,[41] are dependent on extracellular Ca^{2+}. However, some are not, and it seems unlikely that cell Ca^{2+} regulation is the primary target of DTX action.

Although evidence is consistent with DTX being a determinant of virulence in *M. anisopliae*, the precise role(s) of these toxic metabolites is difficult to assess. One approach would be to produce a specific mutant of *M. anisopliae* which is unable to synthesise DTX. This could be achieved by transformation-mediated disruption of a gene essential for toxin biosynthesis, but this first requires that the molecular basis of DTX biosynthesis be elucidated. The majority of microbial peptide antibiotics are synthesised non-ribosomally by the so-called thiotemplate mechanism in a process catalysed by one or more large multifunctional enzymes.[47,48] It has been demonstrated recently that *M. anisopliae* possesses a gene encoding a very large peptide synthetase which shows high identity at the amino acid level to specific conserved regions found in all peptide synthetases.[49] As expected, from the three domains sequenced so far, identity is particularly high with the HC-toxin synthetase of the plant pathogenic fungus *Cochliobolus carbonum*.[50]

Destruxins have been implicated also in plant pathogenesis; DTX B has been reported as the chlorosis-causing principle of *Alternaria brassicae* responsible for black spot in *Brassica* spp.[51] Desmethyl DTX B and the previously unreported homoDTX B were also isolated. DTX B causes necrotic and chlorotic symptoms both on host and nonhost plants, though *Brassica* spp. are more sensitive. It appears to contribute to the aggressiveness of *A. brassicae* by conditioning the host tissue and thereby determining the susceptibility of the host.[52]

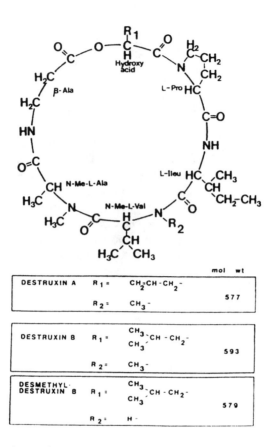

FIGURE 2.   The chemical structure of three major destruxins present in the culture filtrates of *M. anisopliae*. (From Samuels, R. I. et al., *J. Chromatogr. Sci.* 26, 15, 1988. With permission.)

V.   GENETIC VARIATION

Several different techniques have been used recently to investigate the phylogeny of *Metarhizium* spp. and to assess the level of intraspecific genetic variation. These include isozyme analysis as well as a variety of molecular techniques which directly assess polymorphism at the DNA level. Based on morphological (primarily conidiophore and conidium) characteristics, the genus *Metarhizium* is widely recognized as containing three species, *M. anisopliae*, *M. flavoviride* and *M. album*.[53,54] Two varieties of *M. anisopliae* have been described, the short-spored *M. anisopliae* var. *anisopliae* and the long spored var. *majus*.[53] Rombach and co-workers[55] have also recognized two varieties of *M. flavoviride*, var. *flavoviride* and var. *minus*. This classification based on morphology is supported by recent analysis of ribosomal DNA (rDNA) sequence data.[56] A region extending from the 3′ end of the 16S rDNA across the ITS1, the 5.8S rDNA and the ITS2 to the 5′ end of the 28S rDNA was amplified by the polymerase chain reaction (PCR), cloned and then sequenced for 31 isolates of *Metarhizium* spp. Although only two isolates of *M. flavoviride* and one of *M. album* were included in the analysis, these could be distinguished clearly from isolates of *M. anisopliae*. Three New Zealand isolates originally assigned to *M. anisopliae* clustered with the *M. flavoviride/M. album* group. Rakatonirainy and co-workers[57] analyzed the sequences of two variable domains of the 28S rDNA of 41 isolates of *M. anisopliae* and one of *M. flavoviride*. This study also highlighted the distinction between the two species as well as separating *M. anisopliae* var. *anisopliae* from var. *majus*. The phylogenetic tree constructed from rDNA sequences was confirmed by isozyme analysis of a more limited number of isolates. Interestingly, two isolates from New Zealand clustered separately from others of *M. anisopliae*, although not with the one isolate of *M. flavoviride* tested. Analysis of biochemical and molecular characteristics of 24 isolates of *M. anisopliae* and *M. flavoviride* primarily from orthopteran hosts also demonstrated a clear distinction between the two species.[58]

In a study of isozyme variation among isolates of *M. anisopliae* var. *anisopliae* and var. *majus,* Riba and co-workers[59] demonstrated that var. *majus* isolates were relatively homogeneous for isozyme profiles but var. *anisopliae* were highly polymorphic. This observation has been confirmed in a more recent and extensive study.[20] Isozyme variation for eight enzyme systems was investigated for 120 isolates of *Metarhizium* spp. Interestingly, the thirteen isolates of *M. anisopliae* var. *majus* showed multiband phenotypes characteristic of heterozygous diploids, supporting previous suggestions that this variety represents a group of stable diploid isolates.[60] The relative genetic uniformity of var. *majus* isolates apparent in this study is also reflected in their narrow host range, being primarily restricted to scarabid beetles of the genus *Oryctes*.[53,54] Conversely, the substantial genetic polymorphism evident among isolates of var. *anisopliae* is correlated with a very wide host range. Although the study was based on relatively few isolates, St. Leger and co-workers[20] also found a distinction on the basis of isozyme profiles between the isolates of *M. flavoviride* from homopteran hosts and those from other hosts.

Several research groups have investigated variation in *Metarhizium* spp. using the PCR-based technique termed "Random Amplification of Polymorphic DNA" (RAPD).[61–63] Isolates of *M. anisopliae* from several different hosts and from various sugar cane growing regions of Queensland, Australia were analyzed by Fegan and co-workers.[62] A very high degree of polymorphism was evident; 30 distinct RAPD profiles could be distinguished among 31 isolates tested and there was strong geographical clustering of genotypes. Similar high levels of polymorphism were found among isolates of *M. anisopliae* pathogenic towards grasshoppers or locusts,[63] and it was suggested that what is called *M. anisopliae* may actually include a number of cryptic species. Discussion of cryptic species is complicated by the difficulty of defining fungal species, particularly ones for which no sexual stage has been identified.

Genetic variation in *M. anisopliae* is also exhibited in terms of electrophoretic karyotype. Shimizu and co-workers[64] separated chromosomal DNAs by CHEF pulsed-field electrophoresis into seven bands. Using the *Schizosaccharomyces pombe* chromosomes as standards, they estimated chromosome sizes to be between 1.6 to 7.4

megabase pairs with a total genome size of ca. 30 Mbp depending on the isolate. Five isolates tested and analyzed could be distinguished on the basis of their chromosome sizes.

Explaining variation with fungal "species" has to take into account the effects of mutation, recombination, selection and migration. In addition to genomic changes due to point and chromosomal mutations, transposons are being identified in an increasing number of filamentous fungi[65,66] and there is evidence, based on insertions in the nitrate reductase gene, for mobile genetic elements in *M. anisopliae*[67] and *Beauveria bassiana*.[68] Although no sexual cycle has been identified for species of *Metarhizium*, genetic recombination through the parasexual cycle has been demonstrated under laboratory conditions for *M. anisopliae*.[69,70] The significance of parasexuality in natural fungal populations is unclear, but the prospect of the release of nonindigenous or recombinant strains highlights the need for more data. The ease with which isolate-specific molecular markers, e.g., RAPD bands, can be identified should make it possible to address this important issue and provide information on the likelihood of genetic exchange between different genotypes. Molecular markers will also be essential tools for monitoring the fate of novel biocontrol strains released into the environment.

REFERENCES

1. Samson, R. A., Evans, H. C., and Latge, J. P., *Atlas of Entomopathogenic Fungi*, Springer Verlag, Berlin, 1988.
2. Gillespie, A. T., Use of fungi to control pests of agricultural importance in *Fungi in Biological Control Systems*, Burge, M. N., Ed., Manchester University Press, Manchester, U.K., 1988, 37.
3. Bateman, R. et al., The enhanced infectivity of *Metarhizium flavoviride* in oil formulations to desert locusts at low humidities, *Ann. App. Biol.*, 122, 145, 1993.
4. Andersen, S. O., Biochemistry of insect cuticle, *Annu. Rev. Entomol.*, 24, 29, 1979.
5. St. Leger, R. J. et al., *Exp. Mycol.*, 13, 274, 1989.
6. St. Leger, R. J. et al., Prepenetration events during infection of host cuticle by *Metarhizium anisopliae*, *J. Invertebr. Pathol.*, 58, 168, 1991.
7. St. Leger, R. J., Roberts, D. W., and Staples, R. C., Differentiation of appressoria by germlings of *Metarhizium anisopliae*, *J. Invertebr. Pathol.*, 57, 299, 1991.
8. Zhou, X. L. et al., The mechanosensitive channel in whole cells and in membrane patches of the fungus *Uromyces*, *Science*, 253, 1415, 1991.
9. St. Leger, R. J., Roberts, D. W., and Staples, R. C., Novel GTP-binding proteins in plasma membranes of the fungus *Metarhizium anisopliae*, *Biochem. Biophys. Res. Commun.*, 164, 562, 1989.
10. St. Leger, R. J. et al., Protein kinases in the entomopathogenic fungus *Metarhizium anisopliae*, *J. Gen. Microbiol.*, 136, 1401, 1990.
11. St. Leger, R. J. et al., Protein kinases in the entomopathogenic fungus *Metarhizium anisopliae*, *J. Gen. Microbiol.*, 136, 1779, 1990.
12. Lalli, E. and Sassone-Corsi, P., Signal transduction and gene regulation: the nuclear response to cAMP, *J. Biol. Chem.*, 269, 17359, 1994.
13. Screen, S. E. et al., unpublished, 1994.

14. Lee, Y. H. and Dean, R. A., *Plant Cell*, 5, 1993, 693.
15. Dean, R. A. and Mitchell, T. K., *7th ISMPMI abstract*, 42, 120, 1994.
16. St. Leger, R. J., Roberts, D. W., and Staples, R. C., Calcium- and calmodulin-mediated protein synthesis and phosphorylation during germination, growth and protease production by *Metarhizium anisopliae*, *J. Gen. Microbiol.*, 135, 2141, 1989.
17. Bernier, L. et al., *FEMS Microbiol. Lett.*, 60, 261, 1989.
18. St. Leger, R. J. et al., Synthesis of proteins including a cuticle-degrading protease during differentiation of the entomopathogenic fungus *Metarhizium anisopliae*, *Exp. Mycol.*, 13, 253, 1989.
19. St. Leger, R. J., Cooper, R. M., and Charnley, A. K., Production of cuticle-degrading enzymes by the entomopathogen *Metarhizium anisopliae* during infection of cuticles from *Calliphor vomitoria* and *Manduca sexta*, *J. Gen. Microbiol.*, 133, 1371, 1987.
20. St. Leger, R. J. et al., *J. Invertebr. Pathol.*, 60, 89, 1992.
21. Heiler, S., Mendgen, K., and Deising, H., Cellulolytic enzymes of the obligately biotrophic rust fungus *Uromyces viciae-fabae* are regulated differentiation-specifically, *Mycological Res.*, 97, 77, 1993.
22. Goettel, M. S. et al., Ultrastructural localization of a cuticle-degrading protease produced by the entomopathogenic fungus *Metarhizium anisopliae* during penetration of host (*Manduca sexta*) cuticle, *J. Gen. Microbiol.*, 135, 2233, 1989.
23. St. Leger, R. J. et al., Role of extracellular chymoelastase in the virulence of *Metarhizium anisopliae* for *Manduca sexta*, *J. Invertebr. Pathol.*, 52, 285, 1988.
24. St. Leger, R. J., Cooper, R. M., and Charnley, A. K., Distribution of chymoelastases and trypsin-like enzymes in five species of entomopathogenic deuteromycetes, *Arch. Biochem. Biophys.*, 258, 123, 1987.
25. Bidochka, M. J. and Khachatourians, G. G., Purification and properties of an extracellular protease produced by the entomopathogenic fungus *Beauveria bassiana*, *Appl. Environ. Microbiol.*, 53, 1679, 1987.
26. Paterson, I. C. et al., *FEMS Microbiol. Lett.*, 109, 323, 1993.
27. Paterson, I. C. et al., *Microbiology*, 140, 185, 1994.
28. Paterson, I. C. et al., *Microbiology*, 140, 3153, 1994.
29. St. Leger, R. J. et al., Molecular cloning and regulatory analysis of the cuticle-degrading-protease structural gene from the entomopathogenic fungus *Metarhizium anisopliae*, *Eur. J. Biochem.*, 204, 991, 1992.
30. Geremia, R. A. et al., Molecular characterization of the proteinase-encoding gene, *prb1*, related to mycoparasitism by *Trichoderma harzianum*, *Mol. Microbiol.*, 8, 603, 1993.
31. Hassan, A. and Charnley, A. K., *Proc. 10th Int. Congr. Plant. Prot.*, 3, 790, 1983.
32. Hassan. A. and Charnley, A. K., Ultrastructural study of the penetration by *Metarhizium anisopliae* through Dimilin-affected cuticle of *Manduca sexta*, *J. Invertebr. Pathol.*, 54, 117, 1989.
33. St. Leger, R. J., Cooper, R. M., and Charnley, A. K., Cuticle-degrading enzymes of entomopathogenic fungi: synthesis in culture on cuticle, *J. Invertebr. Pathol.*, 48, 85, 1986.
34. St. Leger, R. J., Cooper, R. M., and Charnley, A. K., Cuticle-degrading enzymes of entomopathogenic fungi: degradation in vitro by enzymes from entomopathogens, *J. Invertebr. Pathol.*, 47, 167, 1986.
35. St. Leger, R. J., Staples, R. C., and Roberts, D. W., Entomopathogenic isolates of *Metarhizium anisopliae*, *Beauveria bassiana*, and *Aspergillus flavus* produce multiple extracellular chitinase isozymes, *J. Invertebr. Pathol.*, 61, 81, 1993.

36. St. Leger, R. J., Cooper, R. M., and Charnley, A. K., Cuticle-degrading enzymes of entomopathogenic fungi: regulation of chitinolytic enzymes, *J. Gen. Microbiol.*, 132, 1509, 1986.

37. Roberts, D. W., Toxins of entomopathogenic fungi, in *Microbial Control of Insects, Mites and Plant Diseases*, 2, Burgess, H. D., Ed., Academic Press, New York, 1980, 441.

38. Pais, M., Das, B. C., and Ferron, P., *Phytochemistry*, 20, 715, 1981.

39. Gupta, S., Roberts, D. W., and Renwick, J. A. A., Insecticidal cyclodepsipeptides from *Metarhizium anisopliae*, *J. Chem. Soc., Perkin Trans.*, 1, 2347, 1989.

40. Samuels, R. I., Charnley, A. K., and Reynolds, S. E., The role of destruxins in the pathogenicity of three strains of *Metarhizium anisopliae* for the tobacco hornworm *Manduca sexta*, *Mycopathologia*, 104, 51, 1988.

41. Samuels, R. I., Charnley, A. K., and Reynolds, S. E., *Comp. Biochem. Physiol.*, 90C, 403, 1988.

42. Bradfisch, B. A. and Harmer, S. L., *Toxicon*, 28, 1249, 1990.

43. Jegorov, A., Matha, V., and Hradec, H., *Comp. Biochem. Physiol.*, 103C, 227, 1992.

44. Huxham, I. M., Lackie, A. M., and McCorkindale, N. J., Production of cuticle-degrading enzymes by the entomopathogen *Metarhizium anisopliae* during infection of cuticles from *Calliphora vomitoria* and *Manduca sexta*, *J. Insect Physiol.*, 35, 797, 1989.

45. James, P. J. et al., Inhibition of desert locust (*Schistocerca gregaria*) Malpighian tubule fluid secretion by destruxins, cyclic peptide toxins from the insect pathogenic fungus *Metarhizium anisopliae*, *J. Insect Physiol.*, 39, 97, 1993.

46. Sloman, I. S. and Reynolds, S. E., *Insect Biochem. Mol. Biol.*, 23, 43, 1993.

47. Kershaw, M. J., Reynolds, S. E., and Charnley, A. K., unpublished.

48. Kleinkauf, H. K. and von Döhren, H., Nonribosomal biosynthesis of peptide antibiotics, *Eur. J. Biochem.*, 192, 1, 1990.

49. Kershaw, M. J., Bailey, A., and Clarkson, J. M., unpublished, 1994.

50. Scott-Craig, J. S. et al., The cyclic peptide synthetase catalyzing HC-toxin production in the filamentous fungus *Cochliobolus carbonum* is encoded by a 15.7-kilobase open reading frame, *J. Biol. Chem.*, 267, 26044, 1992.

51. Bains, P. S. and Tewari, J. P., Purification, chemical characterization and host-specificity of the toxin produced by *Alternaria brassicae*, *Physiol. Mol. Plant Pathol.*, 30, 259, 1987.

52. Buchwalt, L. and Green, H., Phytotoxicity of destruxin B and its possible role in the pathogenesis of *Alternaria brassicae*, *Plant Pathol.*, 41, 55, 1992.

53. Tulloch, M., *Trans. Br. Mycol. Soc.*, 66, 407, 1976.

54. Rombach, M. C., Humber, R. A., and Evans, H. C., *Trans. Br. Mycol. Soc.*, 88, 451, 1987.

55. Rombach, M. C., Humber, R. A., and Roberts, D. W., *Mycotaxon*, 27, 87, 1986.

56. Curran, J. et al., Phylogeny of *Metarhizium*: analysis of ribosomal DNA-sequence data, *Mycol. Res.*, 98, 547, 1994.

57. Rakotonirainy, M. S. et al., Phylogenetic-relationships within the genus *Metarhizium* based on 28S RNA sequences and isozyme comparison, M*ycol. Res.*, 98, 225, 1994.

58. Bridge, P. D. et al., Morphological, biochemical and molecular characteristics of *Metarhizium anisopliae* and *M. flavoviride*, *J. Gen. Microbiol.*, 139, 1163, 1993.

59. Riba, G., Bouvier-Fourcade, I., and Caudal, A., Isoenzymes polymorphism in *Metarhizium anisopliae* entomogenous fungi, *Mycopathologia*, 96, 161, 1986.

60. Samuels, K. D. Z., Heale, J. B., and Llewellyn M., Characteristics relating to the pathogenicity of *Metarhizium anisopliae* toward *Nilaparvata lugens*, *J. Invertebr. Pathol.*, 53, 25, 1989.

61. Cobb, B. and Clarkson, J. M., *FEMS Microbiol. Lett.*, 112, 319, 1993.

62. Fegan, M. et al., Random amplified polymorphic DNA markers reveal a high degree of genetic diversity in the entomopathogenic fungus *Metarhizium anisopliae* var. *anisopliae*, *J. Gen. Microbiol.*, 139, 2075, 1993.

63. Bidochka, M. J. et al., Differentiation of species and strains of emtomopathogenic fungi by random amplification of polymorphic DNA (RAPD), *Curr. Genet.*, 25, 107, 1994.

64. Shimizu, S., Arai, Y., and Matsumoto, T., *J. Invertebr. Pathol.*, 60, 185, 1992.

65. Daboussi, M. J., Langin, T., and Brygoo, Y., Fot1: a new family of fungal transposable elements, *Mol. Gen. Genet.*, 232, 12, 1992.

66. Dobinson, K. F., Harris R. E., and Hamer, J. E., *Mol. Plant. Microbe. Interact.*, 6, 114, 1993.

67. Bailey, A. and Clarkson, J. M., unpublished, 1994.

68. Maurer, P., Aioun, A., and Riba, G., *Proc. VIth Int. Coll. Invertebr. Pathol. and Microbial Control*, 2, 319, 1994.

69. Messias, C. and Azevedo, J., *Trans. Br. Mycol. Soc.*, 75, 473, 1980.

70. Al-Aidroos, K., Demonstration of a parasexual cycle in the entomopathogenic fungus *Metarhizium anisopliae*, *Can. J. Genet. Cytol.*, 22, 309, 1980.

71. St. Leger, R. J., Charnley, A. K., and Cooper, R. M., Characterization of cuticle-degrading proteases produced by the entomopathogen *Metarhizium anisopliae*, *Arch. Biochem. Biophys.*, 253, 221,   1987.

72. Cole, S., Cooper, R. M., and Charnley, A. K., *FEMS Microbiol. Lett.*, 1993.

73. St. Leger, R. J., Cooper, R. M., and Charnley, A. K., Analysis of aminopeptidase and dipeptidylpeptidase IV from entomopathogenic fungus *Metarhizium anisopliae*, *J. Gen. Microbiol.*, 139, 237, 1993.

74. St. Leger, R. J., Cooper, R. M., and Charnley, A. K., Cuticle-degrading enzymes of entomopathogenic fungi: regulation of prduction of chitinolytic enzymes, *J. Gen. Microbiol.*, 132, 1509, 1986.

75. St. Leger, R. J., Cooper, R. M., and Charnley, A. K., Characterization of chitinase and chitobiase produced by the entomopathogenic fungus *Metarhizium anisopliae*, *J. Invertebr. Pathol.*, 58, 415, 1991.

9 Molecular Biology of Protozoa for Biological Control of Harmful Insects

Richard A. Heckmann

TABLE OF CONTENTS

I. INTRODUCTION

There is a paucity of information relative to the molecular biology of protozoa successfully used for the control of harmful organisms. Many articles have been published pertaining to those organisms that can be considered for biological control.

0-8493-2442-4/96/$0.00+$.50
© 1996 by CRC Press, Inc.

In our laboratory we first became aware of this control mechanism when *Nosema strigeoidea* was used as a hyperparasite to control the eye fluke, *Diplostomum spathaceum*.[1,2] The molecular mechanism for this response was not investigated, but it was noted that the microsporidan was very effective against the cercariae of *D. spathaceum*. One concern we had was whether or not the microsporidan was host specific. If utilized in aquatic habitats would it infect other organisms including humans?

Biological control is a logical avenue to pursue for control of harmful insects if the biological and ecological implications are understood. Through genetic manipulation it may be feasible to develop more effective biological control agents. To fully understand the biology of these agents, a better database needs to be accumulated for the selected organisms, including the molecular level of control. Harmful insects have become more and more resistant to chemical pesticides; thus the need for other means of control The use of biological control organisms has promise for the future.

The majority of the protozoa that have been successfully used belong to the Phylum Microspora. There is a limited number of manuscripts published on other protozoan groups which can be considered for biological control.

The objective of this paper will be to review the major groups of protozoa, with examples, and emphasize published articles relating to the molecular biology of control.

II.   MICROSPORIDA: MICROSPORA

There are two excellent reviews for this group, those by Weiser[3] and Canning and Lom.[4] The microsporidans have peculiar structures within their cells that have no analogies within other protozoa. Microsporida are small unicellular organisms, all of which are *obligate* intracellular parasites. They are unusual in that they lack mitochondria, thus relying on host cells for chemical energy and in having ribosomes of a size more characteristic of prokaryotic cells. Other cell organelles resemble eukaryotes, including a Golgi complex and a nuclear envelope. Thus the understanding of the molecular mechanism would be more difficult due to these unique structures. As the name connotes, these are small organisms averaging 3–4 μm for the spore stage. The number of spores developed during sporogony is often the basis for classifying the microsporida. The infective agent, known as the sporoplasm, passes from the spore through the tube and enters the cytoplasm of the host cell. In contrast to those parasites which enter host cells in phagosomes, the microsporida multiply in contact with the host cell cytoplasm and are protected from lysosomal attack. They must absorb small molecules from the host cell across their plasma membrane, there being no evidence of pinocytotic activity nor of intracellular digestion of particulate matter in vacuoles.[5,6] Two phases of development are recognised for the microsporida: merogony (also known as schizognony), which is the phase of proliferation, and sporogony, which culminates in the production of sporoblasts. The later undergoes morphogenesis into the highly characteristic spores for transmission to another host cell.

The microsporidans form spores of unicellular origin, with a single sporoplasm discharged from spores through the channel of a tubular polar filament. All active stages of the microsporida develop in host tissues, in the cytoplasm. The invasive stage is a sporoplasm which is a minute uninucleate or binucleate stage which is approximately 1 μm in diamater. The invasive stage is injected into the host's cells through the polar filament by the internal pressure of the spore.[5,6]

Microsporida are parasites of all five classes of vertebrates and many invertebrates and have an impact on biology concerning these hosts. There are numerous species of microsporidans in fish, some of which are responsible for mortality or pathological abnormalities, sufficient to make the fish unmarketable. This manuscript emphasizes the microsporida used as biological control agents for insects.

In some genera, sporogony stages continue to lie directly in the host cell cytoplasm. This is the case for *Nosema*, one of the common microsporidans used for biological control. In other genera, an envelope or covering is produced which isolates or seals off the parasite from the host tissue (*Pleistophora*). The microsporidans are always detrimental to the host cells and in many cases cause major lesions in the tissue which may kill the host. This is the goal for biological control to eliminate the host such as a harmful insect. References exist for the pathogenicity of the microsporidans, but data pertaining to the molecular basis or pathophysiology are very scarce. Describing and understanding the pathological process gives insight to the actual molecular mechanism. Much of this data for the pathological process comes from studies involving fish infected with microsporidans, since this group has had a major effect on the ichthyofauna of the world. Microsporidans may seriously endanger whole stocks of fish and thus reduce productivity. An adequate assessment of their true pathogenic potential even for fish is limited. In aquaculture, new microsporidans emerge which have impact on fish stocks. Microsporidiosis may have the largest impact on young fry and yearling fish and may constitute one of the factors limiting the growth of stocks, which is similar to listed biological control agents attacking larval stages of the insect. Following is a list of examples used for biological control of insects.

A. *NOSEMA*: (NAEGELI 1857)

Nosema: (Naegeli 1857): Nosematidae: Labbé 1899

Characteristics of this genus include; oval shape spores, sporoblast with a diplokaryon (two nuclei per spore), with sporogony by binary fission of the sporont; all examples are intracellular and parasitic. This is a common genus of microsporidan used for biological control.

Perezia:[3,7,8] (Léger and Duboscq 1909): Glugeidae Thélohan 1892

Disporoblastic organism. Nuclei unpaired up to the spore stage. Infected cells do not hypertrophy, without tumor formation. Often considered *Nosema*, but *Perezia* has unpaired nuclei.

The genus *Nosema* is one of the major groups of protozoa used for biological control. Being intracellular, the organism redirects the metabolism of host cells to form more *Nosema*, similar to the effect viruses have on host cells. For *N. locustae*, the microorganism invades the fat bodies and pericardium of the grasshopper,[9,10,11] whereby energy for regenerating more cells may be obtained for the parasite. One of the primary goals for biological control is to negate the use of chemical agents or use the chemical compound (insecticides) in combination with the biological control organism. The use of the genus *Nosema* has had varied results, which is the case for many biological control agents. In some cases no response is noted, as exemplified by Kantuck's[12] paper, while other authors have recorded success with *Nosema*.[11] Perhaps a combination of biological control organisms in which results are achieved creates problems in understanding the molecular basis.

Some of the examples of species of *Nosema* utilized for biological control are listed in Table 1.

TABLE 1.

Examples of the Genus *Nosema* Considered for Biological Control of Harmful Insects and Pathogens

	Parasite	Host	References
1.	Nosema acridophagus	Grasshoppers	Maugh[13]
2.	Nosema algerae	Mosquitoes	Akhtar et al.[14]
		Anopheles albimanus	Avery and Anthony[15]
		Culex tritaeniorhynchus	Canning and Hull[16]
			Lai and Canning[17]
			Maddox et al.[18]
			Sheffield et al.[19]
			Undeen[20]
			Undeen and Alger[21,22]
			Undeen and Maddox[23]
			Van Essen and Anthony[24]
3.	Nosema apis	Honey bees	Menapace et al.[25]
		Apis melanifera	
4.	Nosema cuneatum	Grasshoppers	Henry and Oma[10]
		Melanoplus bivittatus	
5.	Nosema disstriae	Forest tent caterpillar	Wilson[26]
		Melacosoma disstriae	
6.	Nosema fumiferanae	Spruce budworm	Bauer and Nordin[27]
		Choristoneura fumiferanae	McMorran[28]
			Nealis and Smith[29]
			Sanders and Wilson[30]
			Thompson[31]
			Wilson[32–35]
		Malacocoma pulviale	Wilson[36]
7.	Nosema furnacalis	Ostrinia furnacalis	Munderloh et al.[37]

TABLE 1 (CONTINUED).

Parasite	Host	References
8. *Nosema heliothidis*	Corn earworm *Heliothis zea*	Brooks et al.[38] Gaugler and Brooks[39] Hamm et al.[40]
Nosema necatrix	Army worm *Pseudaletia unipuncta*	Kramer[41]
9. *Nosema locustae*	Grasshoppers several species (58+) *Locusta migratoria* Desert locust	Germida[42] Henry [43–45] Henry and Oma[9–11] Kantack[12] Kramer[41] Lockwood and Debrey[46] Srivastaya and Bhanotar[47] Strett and Henry[48]
	Mormon crickets *Anabrus simplex* Silkworm *Bombyx mori*	Henry and Onsager[49] MacVean and Capinera[50] Raina et al.[51]
10. *Nosema pyrausta*	European corn borer *Ostrinia nubilalis*	Cossentine and Lewis[52] Kramer[41] Lewis[53] Lewis et al.[54,55] Zimmack et al.[7,8]
	Braconid *Macrocentrus grandii* Neuropteran *Chrysoperla carnae*	Speigel et al.[56] Sajap and Lewis[57]
11. *Nosema whitei*	Spruce budworm *Choristoneura fumiferanae*	Wilson and Sohi[58] Thomson[31]
12. *Nosema wistmansi*	Winter moths *Operophtera brumata*	Canning et al.[59]

B.   *PLEISTOPHORA*: (GURLEY 1893)

Pleistophora: (Gurley 1893): Pleistophoridae (Stempell 1909)

The major characteristics of this organism are uninucleate bodies without envelopes, and the vegetative stage develops into a plasmodium which is often amoeboid. The spores are 3 μm × 1.5 μm and have an ovoid shape. They are intracellular parasites which are commonly found in fish.[3,6,60]

Examples of those used for the biological control of harmful insects are found in Table 2.

TABLE 2.

Examples of the Genus *Pleistophora* Considered for Biological Control of Harmful Insects and Pathogens

Parasite	Host	References
1. *Pleistophora operophtera*	Winter moths *Operophtera brumata*	Canning et al.[59]
2. *Pleistophora schubergi*	Geometrid moth *Anaitas efformate* Spruce budworm *Choristoneura fumiferana*	Bulla and Cheng[60] Milner and Briese[61] Wilson[32–35,62]

C. *VAIRIMORPHA*: (PILLEY 1976)

Vairimorpha: (Pilley 1976)[63] Nosematidae: Labbé 1899

This genus was proposed for a dimorphic species for *Nosema necatrix*.[6,60] This name supersedes *Microsporidium necatrix*. Thus, the genus should be included with the *Nosema*. The genus *Vairimorpha* has been used for biological control, with examples listed in Table 3.

TABLE 3.

Examples of the Genus *Vairimorpha* Considered for Biological Control of Harmful Insects and Pathogens

Parasite	Host	References
1. *Vairimorpha necatrix*	Black cutworm larvae *Agrotis ipsilon*	Cossentine and Lewis[64] Kurti et al.[65] Grundler et al.[66]
	Colorado potato beetle *Leptinotarsa decemlineata*	Jaques and Laing[67]
	Tachinid *Bonnetia comta*	Cossentine and Lewis[52,68]
	Corn earworm *Heliothis zea*	Mitchell and Cali[69]

D. *GLUGEA*: (THÉLOHAN 1891)

Glugea: (Thélohan, 1891): Glugeidae: (Thélohan 1892)

These eucaryotic cells are intracellular parasites with sporogony in a membrane bound vacuole. Diplosporoblastic sporonts arise by plasmotomy of a the plasmodial stage. The infected cells enlarge (hypertrophy), forming a tumor or xenoma.[3,6]

Examples of those used for biological control of harmful insects are found in Table 4.

TABLE 4.

Examples of the Genus *Glugea* Considered for Biological Control of Harmful Insects and Pathogens

	Parasite	Host	References
1.	*Glugea gasti*	Boll Weevil	McLaughlin et al.[70,71]
		Anthonomus grandis	McLaughlin[72,73]

III. OTHER PROTOZOA

Representatives of the eugregarines and amoeba have also been used as biological control agents.[74] These are two to three times the size of the microsporidans. Examples of these protozoa are found in Table 5.

TABLE 5.

Examples of the Other Protozoa Considered for Biological Control of Harmful Insects and Pathogens

	Parasite	Host	References
1.	*Amblyospora californica*	Mosquito	Samson et al.[75]
		Culex tarsalis	
2.	*Ascogregarina culicis*	Mosquito	Spencer and Olson[76]
		Aedes	
		Aedes aegypti	Suliaman[77]
3.	*Gregarina garnham*	Mosquito	Harry[78]
			Harry and Finlayson[79]
4.	*Malameba locustae*	Grasshoppers	Braun et al.[80]
		Melanoplus sanguinipes	Hanrahan[81,82]
			Harry and Finlayson[79]
			Henry[83]
5.	*Malpighamoeba locustae*	Grasshoppers	King and Taylor[84]
6.	*Mattesia grandis*	Boll weevil	McLaughlin[72]
		Anthonomous grandis	McLaughlin et al.[70]

A. ASCOGREGARINA

Ascogregarina culicis has been suggested as a biological control agent for *Aedes* mosquitoes.[77] Mosquito larvae become infected through ingestion of the parasite oocyst. The trophozoite stage of the parasite develops in the epithelial cells of the digestive tract of the mosquito larva. This stage later occupies the space between the epithelial lining of the gut and the peritrophic membrane as large gamonts. Following host pupation, the gamonts migrate to the distal ends of the Malpighian tubules and fuse in pairs to form gametocysts. Numerous oocysts develop in each gametocyst. Each oocyst contains eight sporozoites which are released when the mosquito defecates or dies.[77]

Reports on the pathogenicity of ascogregarine parasites to mosquito larvae have been diverse and contradictory. Results of the Suliaman study[77] indicated that *A. culicis* has some potential as a biological control agent for *Aedes aegypti*. Because strains of *A. culicis* differ in pathogenicity, it is likely that there are natural strains more pathogenic than those studied in this research. One amoeboid protozoan has shown promise as a biological control agent of grasshoppers and locusts.[80] Infected insects are sluggish and have reduced nutrition and fecundity.

IV. BIOLOGY OF THE MICROSPORA

Bulla and Cheng[5,60] review both the systematics and biology of the microsporidans. There has been an increase in the number of papers published on the biological control of harmful insects due to interest in pest management. Henry[85] published an excellent review on the use of protozoa for biological control. Again, there is a paucity of information in his review on the molecular aspects of the host–parasite interaction. An all-inclusive publication for the microbial control of vectors was authored by Davidson and Sweeney.[86] The latter review emphasized recent progress in the microbial arena, listing those organisms of current interest which were primarily effective against mosquito larvae, grasshoppers, budworms and black fly larvae. There is a large group of potentially useful organisms in nature that have yet to be investigated.

To take maximum advantage of the potential of these organisms, new techniques for their culture, preservation, formulation, and application must be developed, and their production must be made cost-effective. Advanced technology, such as genetic manipulaiton, is already being applied to these microorganisms, which should result in enhanced efficacy and host range. Future integrated control programs for insect vectors can be expected to contain an important microbial control component.[86] As research goals are developed for these control candidates, emphasis should be placed on the biological mechanism involved in the death of the vectors, thus clarifying the molecular interaction between the host and parasite.

A combination of microbials and chemical insecticides represents one future for control of harmful insects.[14,39,50,55,64,68,85] Current research is showing increased efficacy when protozoa are applied in baits or on particulate carriers. Development of new formulation and application techniques is necessary for delivery to the vector. Because of the characteristically low virulence expressed by most protozoa, especially the microsporida, few, if any, will be used alone against insect pests. They will be integrated with other control agents. Microbials have generally been evaluated against standards for relatively inexpensive, broad spectrum chemical insecticides.

Against such standards, few microbials could compete.[85] There has been a significant increase in integrated pest management programs in which protozoa and other microbials can and will play a role. When a combination of insecticides and microbial agents are used, it will be even more difficult to understand the molecular basis of control for the vector. As these agents are developed, what is the effect on beneficial organisms? Do they adversely affect humans? Research involving *Nosema locustae* has shown that the microsporidan does not harm honey bees.[25] Sajap and Lewis[87] have also published articles relative to the effect of the microbial agent to beneficial organisms. Some authors foresee the end of insect plagues through using microbial control and insecticides.[13] Mass production[17,32,45,88,89] and application[12,43,46,90] represent the applied aspects of employing these pathogens by studying the transmission within the microhabitat.[44,45,53,54,57,69,79,91]

There is a world wide effort in trying to control harmful organisms with biological control. This is exemplified by Raina et al.[51] who outline an effort to present a new approach to grasshopper and locust control in India through the use of microsporidans. The objective of the authors is to replace hazardous and uneconomical chemical pesticides with inexpensive and ecologically safe biocides of protozoan origin. The protozoan *N. locustae* was used locally and developed in an insectary, using locusts. It was screened against the silkworm to make it acceptable for field use.

V.   MOLECULAR CHARACTERISTICS OF CONTROL ORGANISMS

The molecular characteristics of *Nosema* have been studied and shed light on the control basis for the vectors.[48] Polypeptide profiles of the spore structure of *N. locustae* have been studied with the best profiles for both the exospore and spore polypeptides following treatment with buffers that contain 4% sodium dodecyl sulfate and 0.02 M dithiothreitol.[48]

Another research aspect has been the characterization of the biological control agent with electron optics.[15] This approach should be used for the host–parasite interaction which would shed light on the molecular basis for control. Cell culture can be used to understand more about the biology of the microbial control agent. An excellent review for the response of microsporida in cell culture is by Jaronski.[92]

In vitro culture of Microsporida with continuous cell lines is feasible. The major requirements for success are a sufficient number of aseptic spores, the appropriate germination stimuli, availability of cell lines, and diligent investigators.[92] Recently, biochemical studies have included comparison of chromosomal DNA for different species of microsporidans to see if the species names are valid.[37]

Spores of two microsporida, *Nosema pyrausta* (from the European corn borer, *Ostrinia nubilalis*) and *N. furnacalis* (from the Asian corn borer, *O. furnacalis*), were harvested from laboratory-reared *O. nubilalis* caterpillars and purified by centrifugation through Percoll. Conditions permitting in vitro germination were defined for both species and found to be different. This study[37] may suggest that strains of *N. pyrausta* and *N. furnacalis* are more closely related than previously suspected. Better detection methods are being used to determine the presence of the microbial agent in the vector and longivity within the microhabitat. These methods include the specific enzyme-linked immnosorbert assays (ELISHA).[93]

Bauer and Nordin[94] studied the nutritional physiology of *Choristoneura fumiferana* infected with the microbial control agent, *Nosema fumiferana*. Nutritional deficiencies of the host would give insight to the interaction of the two organisms and data for the molecular basis. Results of their study were as follows:

> Female eastern spruce budworm larvae, *Choristoneura fumiferana* (Clemens) (Lepidoptera: Tortricidae), inoculated with a medium lethal spore dosage of the microspordium *Nosema fumiferanae* (Thomson) exhibited significant reductions in consumptive index (CI), nitrogen consumptive index (NCI), relative growth rate (RGR), and gross (ECI) and net (ECD) production efficiencies when compared to microsporidan-free larvae. Diseased larvae also exhibited significant increases in approximate digestibility (AD), N utilization efficiency (NUE), and larval moisture content. Both healthy and diseased insects were reared on 2.5% N and 4.5% N diets. Those on the 2.5% N diet showed significant increase in CI, although NCI was still lower than NCI measured for larvae reared on 4.5% N. NUE was also higher on the 2.5% N diet. Diseased cohorts reared on 2.5% N diet had significantly greater mortality than those reared on 4.5% N diet. Pupal weight and development time of infected individuals did not respond to dietary N concentration. However, healthy insects achieved greater pupal weights in a shorter time on the 4.5% N diet than those on the 2.5% N diet. Mortality of healthy insects was unaffected by dietary N.

The nitrogen deficiency in this study substantiates the redirecting of host cell metabolism for the production and metabolic needs of the parasite. It is feasible that elevated levels of nitrogen will improve the growth and survival of both the microbial control agent and the host.

VI. EFFECT ON HOST CELLS: MOLECULAR LEVEL

As indicated, microsporidans are intracellular parasites—thus the molecular basis for their host response is at the cell level.

The growth and division of the microsporidan within the host cell always results in the complete destruction of the latter. There are a variety of interactions between the parasite and the host cell. In the genus *Pleistophora* (one listed in this article), developmental stages and, later, mature spores, gradually replace the cell contents until the host cell (or syncytium in muscle fibers) becomes a mere envelope—sometimes slightly enlarged—containing the parasite. This indicates lytic action, which turns the cytoplasm into an unorganized mass containing an array of vesicles, endoplasmic reticulum cisternae and ribosomes.[5,86] The cell structure beyond the infected region may not necessarily reveal pathological changes. In other genera, the microsporidan stimulates the infected cell to an enormous hypertrophy. Such hypertrophic cells—they are usually connective tissue cells or cells of mesenchyme origin—can reach dimensions up to 14 mm and are called xenomas.[3,85] In the xenoma, the host cell, with a completely changed structure, and its microsporidan parasite, are physiologically and morphologically integrated to form a separate entity with its own development in the organism of the host, at the expense of which it grows.

The xenomas proliferate and cause inflammation, its onset being linked to changes in the xenoma wall. The newly formed granulation tissue invades the xenoma contents, turning the xenoma into a granuloma with a gradually diminishing spore mass in the center.[6]

The tissue reaction in xenoma-infection is, from its first stage, directed toward the isolation of the parasite and very often results in a complete repair. The possibility of auto-infection seems to be very limited; secondary xenomas produced by spores from within the xenoma were never found outside its confines.

The ultimate fate of all spores produced by any kind of microsporidan infection is the ingestion by phagocytes, unless they are discharged to the outer milieu or unless the host perishes while they are still viable. The later is the response generated by acceptable biological control agents. Phagocytic cells of the granulation tissue, including the fibroblasts, play a pivotal role in the host defense mechanism; they may destroy the spores on the spot or carry them further away. Spore-filled macrophages in the granuloma may themselves disintegrate, and the resulting spore-containing debris is taken up by other phagocytes. Within the phagocytic cell, spore contents are digested through the spore shell, which then appears empty and disintegrates until it is reduced to a thin folded membrane, which later disappears completely. This is evidence of the considerable enzymatic capability of the phagocyte, including a chitinase. There is no evidence that phagocytized spores can extrude and initiate a new infection.

Our understanding of microsporidan physiology and biochemistry is definitely inadequate. Much information about microsporidan metabolism can be revealed by utilizing radiolabeled compounds and autoradiography. Analyses of cell fractions by isoelectric focusing and polyacrylamide gel electrophoreis of microsporidian enzymes, high performance liqid chromatography of normal and labeled metabolic intermediates, and ultracytochemical localization of specific enzymes can uncover much about the metabolism of microsporida.[92] The microsporida examined have no mitochondria; thus host cell mitochondria and endoplasmic reticulum aggregate near the parasites, which probably redirects the energy cycle of the host cell to the microsporidan. Nothing is known about the importance of the Krebs cycle or the Embden—Meyerhof pathway in the microsporida.

We currently have no information about the biochemical events that create the visible microsporidan architecture or about the biochemistry of morphological differentiation.[92]

Pathobiological studies have generally been restricted to histological observations and measurements of metabolic changes in intact hosts. Nothing is known about the metabolic changes within infected cells as the microsporida develop. Authors have suggested that the microsporidium acts like a mutualist during merogony by stimulating host-cell protein synthesis, growth, and endomitosis. During sporogony, the microsporidium walls itself off from the host cell and behaves like a pathogen, killing and lysing the cell.[5,92]

VII. CONCLUSIONS

There are numerous protozoa of the microsporidan group available for biological control of harmful insects. The primary genus for this group is *Nosema*, with species effective against grasshoppers (*N. locustae*), mosquitoes (*N. algarae*), spruce budworm (*N. fumiferanae*), corn earworm (*N. heliothidis*) and european corn borer (*N. pyrausta*). Application methods, transmission, life cycles and the biology of these species are well understood, but the host cell response has limited data. To understand the molecular basis of control it is necessary to infer these responses without an adequate database. All microspiridans are intracellular parasites with metabolism of the host cell directed towards forming more microsporidans. For *Glugea* infected cells, there is a marked hypertrophy of the target tissue forming a tumor or xenoma. Microspiridans are characterized by their size (2–5 μm for spores), with an absence of mitochondria in the organism. Other protozoa used for biological control include the eugregarina and sarcodina.

The most recent studies center on detection methods, combining insecticides with microbials, application methods and transmission. Knowledge of the biochemical pathways within the microsporidan is limited with nucleic acid studies and protein profiles being published. There has been a worldwide effort to try and decrease the use of chemical insecticides and to use biological control agents. With this current interest more papers will be published in the future pertaining to the molecular basis of biological control. Better detection methods are being used to determine the presence of the microbial agent. It will be a challenge for future research to establish a good data base for the molecular basis of biological control.

REFERENCES

1. Palmieri, J. R., Cali, A., and Heckmann, R. A., Experimental biological control of the eye fluke *Diplostomum spathaceum*, by a protozoan hyperparasite *Nosema strigeoidea* (Protozoa: Microsporida). *J. Parasitol.*, 62, 325, 1976.
2. Shingina, N. G. and Grobov, O. F., *Nosema diplostomi* sp.n. (Microsporidia: Nosematidae), a hyperparasite of trematode of the genus *Diplostomum*. *Parazitologiya* (Leningrad), 6, 269, 1972.
3. Weiser, J., Phylum Microsporda Sprague, in *Illustrated Guide to the Protozoa*, Lee, J. J., Hutner S. H., and Boree E. C., Eds. Allen Press, Lawrence, KS, 1969, 375–383.
4. Canning, E. U. and Lom, J., *The Microsporidans of Vertebrates*. Academic Press, New York, 1986.
5. Bulla, L. A., Jr. and Cheng, T. C., *Comparative Pathobiology*, 1. *Biology of the microsporidia*. Plenum Press, New York, 1976.

6. Lom, J. and DyKova, I., *Protozoan Parasites of Fishes. Developments in Aquaculture and Fisheries Science*, 26. Elsevier Publishers, New York, 1992, 125–154.

7. Zimmack, H. L. and Brindley, T. A., The effect of the protozoan parasite *Perezia pyraustae* Paillot on the European corn borer. *J. Econ. Entomol.*, 50, 637, 1957.

8. Zimmack, H. L., Arbuthnot, K. D., and Brindley, T. A., Distribution of the European corn borer parasite *Perezia pyrausta*, and its effect on the host. *J. Econ. Entomol.*, 47, 641, 1954.

9. Henry, J. E. and Oma, E. A., Effect of prolonged storage of spores on field applications of *Nosema locustae* (Microsporida: Nosematidae) against grasshoppers. *J. Invert. Pathol.*, 23, 271, 1974.

10. Henry, J. E. and Oma, E. A., Effects of infections by *Nosema locustae* Canning, *Nosema acadophagus* Henry, and *Nosema cuneatum* Henry (Microsporida: Nosematidae) in *Melanoplus bivitattus* (Say) (Orthoptera: Acrididae). *Acrida*, 3, 223, 1974.

11. Henry J. E. and Oma, E. A., Pest control by *Nosema locustae*, a pathogen of grasshoppers and crickets, in *Microbial Control of Pests and Plant Diseases*, Burges, H. D., Ed., Academic Press, New York, 573, 1981.

12. Kantack, K., Field Facts. Cooperative Extension Service, South Dakota State University, 3, 2, 1988.

13. Maugh, J. H., II, The day of the locust is near. *Science*, 313, 1981.

14. Akhtar, R., Hayes, C. G., and Bagar S., Dual infections of *Culex triaeniorhynchus* with west Nile virus and *Nosema algerae*. *J. Parasitol.*, 67:571, 1981.

15. Avery, S. W. and.Anthony D. W., Ultrastructural study of early development of *Nosema algerae* in *Arropheles albimanus*. *J. Invert. Path.*, 42, 87, 1983.

16. Canning, E. U. and Hull, R. J., A microsporidian infection of *Anopheles gambiae* Giles from Tanzania, interpretation of its mode of transmission and notes on *Nosema* infections in mosquitoes. *J. Protozool.*, 17, 531, 1970.

17. Lai, P. F. and Canning, E. U., Infectivity of a microsporidium of mosquitoes (*Nosema algerae*) to larval stages of *Schistosoma mansoni* in *Biomphalaria glabrata*. *Int. J. of Parasitol.*, 10, 293, 1980.

18. Maddox, J. V., Alger, N. E., Allmad, A., and Aslamkiian, M., The susceptibility of some Pakistan mosquitoes to *Nosema algerae* (Microsporidia) *Pakistan J. Zool.*, 9, 19, 1977.

19. Sheffield, H. G., Garnham, P. C. C., and Shioishi, T., The fine structure of the sporozoite of *Lankesteria culicis*. *J. Protozool.*, 18, 98, 1971.

20. Undeen, A. H., Growth of *Nosema algerae* in pig kidney cell cultures. *J. Protozool.*, 22, 107, 1975.

21. Undeen, A. H. and Alger, N. E., The effect of the microsporidian, *Nosema algerae*, on *Anopheles stephensi*. *J. Invertebr. Pathol.*, 25, 19, 1975.

22. Undeen, A. H. and Alger, N. E., *Nosema algerae*: infection of the white mouse by a mosquito parasite. *Exp. Parasitol.*, 40, 86, 1976.

23. Undeen, A. H. and Maddox, J. V., The infection of nonmosquito hosts by injection with spores of the microsporidan *Nosema algerae*. *J. Invert. Pathol.*, 22, 258, 1973.

24. Van Essen, F. W. and Anthony, D. W., Susceptibility of nontarget organisms to *Nosema algerae* (Microsporida: Nosematidae), a parasite of mosquitoes. *J. Invert. Pathol.*, 23, 77, 1976.

25. Menapace, D. M., Sackett, R. R., and Wilson, W. T., Adult honey bees are not susceptible to infection by *Nosema locustae*. *J. Econ. Entomol.*, 71, 304, 1978.

26. Wilson, G. G., Pathogenicity of *Nosema disstriae, Pleistophora schubergi* and *Vairimorpha necatrix* (Microsporidia) to larvae of the forest tent caterpillar, *Malacosoma disstria*. *Z. Parasitenkd.*, 70, 763, 1984.

27. Bauer, L. S. and Nordin G. L., Response of the Spruce budworm (Lepidoptera: Tortricidae) infected with *Nosema fumiferanae* (Microsporida) to *Bacillus thuringiensis* treatments. *Environ. Entomol.*, 18, 816, 1989.

28. McMorran, A. R., A synthetic diet for the spruce budworm, *Choristoneura fumiferana* (Clem.), (Lepidoptera: Tortricidae). *Can. Entomol.*, 97, 58, 1965.

29. Nealis, V. G. and Smith, S. M., Interaction of *Apanteles fumiferana* (Hymenoptera: Braconidae) and *Nosema* fumiferana (Microsporidia) parasitizing the spruce budworm, *Choristoneura fumiferana* (Lepidoptera: Tortricidae). *Can. J. Zool.*, 65, 2047, 1987.

30. Sanders, C. J. and Wilson, G. G., Flight duration of male spruce budworm (*Choristoneura fumiferana* Cem.) and attractiveness of female spruce budworm are unaffected by microsporidan infection or moth size. *Can. Entomol.*, 122, 419, 1990.

31. Thompson, H. M., The effect of a microspiridian parasite on the development, reproduction and mortality of the spruce budworm, *Choristoneura fumiferana* (Clem.). *Can. J. Zool.*, 36, 499, 1958.

32. Wilson, G. G., A method for mass producing spores of the microsporidan *Nosema fumiferanae* in its host, the spruce budworm, *Choristoneura fumiferanae* (Lepidoptera: Tortricidae). *Can. Entomol.*, 108, 383, 1976.

33. Wilson, G. G., Effects of *Nosema fumiferanae* (Microsporida) on the rearing stock of spruce budworm *Choristoneura fumiferana* (Lipidoptera: Tortricidae). *Proc. Entomol. Soc. Ontario*, 111, 115, 1980.

34. Wilson, G. G., Transmission of *Nosema fumiferanae* (Microsporida) to its host *Choristoneura fumiferanae*. *Z. Parasitenkd.*, 68, 47, 1982.

35. Wilson, G. G., Transmission of *Nosema fumiferanae* (Microsporida) to its host *Choristoneura fumiferanae*. *Z. Parasitenkd.*, 68, 47, 1982.

36. Wilson, G. G., The effects of termperature and ultraviolet radiation on the infection of *Choristoneura fumiferanae* and *Malacosoma pluviale* by the microsporidan parasite, *Nosema (Perezia) fumiferanae*. *Can. J. Zool.*, 52, 59, 1974.

37. Munderloh, V. G., Kuritti, T. J., and Ross, S. E., Electrophoretic characterizations of chromosomal DNA from two microsporida. *J. Invert. Pathol.*, 55, 243, 1990.

38. Brooks, W. M., Crawford, J. D., and Pearce, L. W., Benomyl: Effectiveness against a microsporidian *Nosema heliothidis* in the corn earworm, *Heliothis zea. J. Invertebr. Pathol.*, 31, 239, 1978.

39. Gaugler, R. R. and Brooks, W. M., Sublethal effects of infection by *Nosema heliothidis* in the corn earworm, *Heliothis zea. J. Invertebr. Pathol.*, 26, 57, 1974.

40. Hamm, J. J., Burton, R. L., Young, J. R., and Daniel, R. T., Elimination of *Nosema heliothidis* from a laboratory colony of the corn earworm. *Ann. Entomol. Soc. Am.*, 64, 624, 1971.

41. Kramer, J. P., *Nosema necatrix* sp.n. and *Thelohania diazoma* sp. n. Microsporidians from the armyworm *Pseudaletia unipuncta* (Haworth). *J. Invert. Pathol.*, 7, 117, 1965.

42. Germida, J. J., Persistence of *Nosema locustae* spores in soil as determined by fluorescence microscopy. *Appl. Environ. Microbiol.*, 47, 313, 1984.

43. Henry, J. E., Results of the 1969 field applications of *Nosema locustae* for grasshopper control. *USDA-ARS, Special Report* Z-230, 1970.

44. Henry, J. E., Experimental application of *Nosema locustae* for control of grasshoppers. *J. Invert. Pathol.*, 18, 389, 1971.

45. Henry, J. E., Production and commercialization of microbials. . . *Nosema locustae* and other protozoa in *Proc. 3rd Int. Colloq. Invertebrate Pathology*, 1992, 103.

46. Lockwood, J. A. and Debrey, L. D., Direct and indirect effects of a large-scale application of *Nosema locustae* (Microsporida: Nosematidae) on rangeland grasshoppers (Orthoptera: Acrididae). *J. Econ. Entomol.*, 83, 377, 1990.

47. Scrivastava, Y. N. and Bhanotar, R. K., A new record of a protozoan pathogen *N. locustae* Canning infecting desert locust from Rajasthan. *Indian J. Entomol.*, 45, 500, 1983.

48. Strett, D. A. and Henry, J. E., Effect of various sodium dodecyl sulfate sample buffers on the exospore and spore polypeptide profiles of *Nosema locustae* (Microspora: Nosematidae). *J. Parasitol.*, 71, 831, 1985.

49. Henry, J. E. and Onsager, J. A., Experimental control of the Mormon cricket, *Anabrus simplex*, by *Nosema locustae* (Microspora: Microsporida) a protozoan parasite of grasshoppers (Ort.: Acrididae). *Entomophaga*, 27, 197, 1982.

50. MacVean, C. M. and Capinera, J. L., Field evaluation of two microsporidan pathogens, an entomopathogenic nematode, and Carbaryl for suppression of the Mormon cricket, *Anabrus simplex* (Orthoptera: Tettigoniidae). *Biological Control* 2, 59, 1992.

51. Raina, S. K., Rai, M. M., and Khurad, A. M., Grasshopper and locust control using microsporidian insecticides. In *Biotechnology in Invertebrate Pathology and Cell Culture*, Academic Press, New York, 1987, 127.

52. Cossentine, J. E. and Lewis, L. C., Impact of *Nosema pyrausta, Nosema* sp. and a nuclear polyhedrosis virus on *Lydella thompsoni* within infected *Ostrinia nubilalis. J. Invert. Pathol.*, 651, 126, 1988.

53. Lewis, L. C., Persistance of *Nosema pyrausta* and *Vairimorpha necatrix* measured by microsporidiosis in the european corn borer. *J. Econ. Entomol.*, 75, 670, 1982.

54. Lewis, L. C., Cossentine, J. E., and Gunnarson, R. D., Impact of two microspirida, *Nosema pyrausta* and *Vairimorpha nexatrix*, in *Nosema pyrausta* infected European corn borer (*Ostrinia nubilalis*) larvae. *Can. J. Zool.*, 61, 915, 1983.

55. Lewis, L. C., Lublinkhof, J., Berry, E. C., and Gunnarson, R. D., Response of *Ostrinia nubilalis* (Lep., Pyralidae) infected with *Nosema pyrausta* (Microsporida: Nosematidae) to insecticides. *Entomophaga*, 27, 211, 1982.

56. Spiegel, J. P., Maddox, J. V., and Ruesink, W.G., Impact of *Nosema pyrausta* on a braconid, *Macrocentrus grandii*, in Central Illinois. *J. Invert. Pathol.*, 47, 271, 1986.

57. Sajap, A. S. and Lewis, L. C., Chronology of infection of European corn borer (Lepidoptera: Pyralidae) with the microsporidium *Nosema pyrausta*: effect on development and vertical transmission. *Environ. Entomol.*, 21, 178, 1992.

58. Wilson, G. G. and Sohi, S. S., Inoculation of the spruce budworm cell cultures with the microsporidium, *Nosemi whitei*. Report FPM-X-24, Canadian Forestry Service, Sault Ste. Marie, Canada, 1979.

59. Canning, E. U., Wigley, P. J., and Barker, R. J., The taxonomy of three species of microsporidia (Protozoa: Microspora) from an oakwood population of winter moths *Operophtera brumata* (L.) (Lepid optera: Geometridae). *Syst. Parasitol.*, 5, 147, 1983.

60. Bulla, L. A., Jr. and Cheng, T. C., *Comparative Pathobiology*, Vol. 2. *Systematics of the Microsporidia*. Plenum Press, New York, 1977.

61. Milner, R. J. and Briese, D. T., Identification of the microsporidan *Pleistophora schugergi* infecting *Anaitis efformata* (Leidoptera: Geometridae). *J. Invert. Pathol.*, 48, 100, 1986.

62. Wilson, G. G., Effects of *Pleistophora schubergi* (Microsporida) on the spruce budworm, *Choristoneura fumiferana* (Lepidoptera: Tortricidae). *Can. Entomol.*, 114, 81, 1982.

63. Pilley, B. M., A new genus, *Vairimorpha* (Protozoa: Microsporida) for *Nosema necatrix* Kramer 1965: pathogenicity and life cycle in *Spodoptera exempta* (Lepidoptera: Noctuidae). *J. Invert. Pathol.*, 28, 177, 1976.

64. Cossentine, J. E. and Lewis, L. C., Interactions between *Vairimorpha nexatrix, Vairimorpha* sp. and a nuclear polyhedrosis virus from *Rachipulsia ou* in *Agrotis ipsilon* larvae. *J. Invert. Pathol.*, 28, 35, 1984.

65. Kurtti, T. J., Munderloh, U. G., and Noda, H., *Vairimorpha nexatrix:* infectivity for and development in a lepidopteran cell line. *J. Invert. Pathol.*, 55, 61, 1990.

66. Grundler, J. A., Hostetter, D. L., and Keaster, A. J., Laboratory evaluation of *Vairimorpha nexatrix* (Microspora: Microsporidia) as a control agent for the black cutworm (Lepidoptora: Noctuidae). *Environ. Entomol.*, 16, 1228, 1987.

67. Jaques, R. P. and Laing, D. R., Effectiveness of microbial and chemical insecticides in control of the Colorado potato beetle (Coleoptera: Chrysomelidae) on potatoes and tomatoes. *Can. Entomol.*, 120, 1123, 1988.

68. Cossentine, J. E. and Lewis, L. C., Impact of *Vairimorpha nexatrix* and *Vairimorpha* sp. (Microspora: Microsporida) on *Bonnetia comta* (Diptera: Tachinidae) within *Agrothis ipsilon* (Lepidoptera: Noctuidae) hosts. *J. Invert. Pathol.*, 47, 303, 1986.

69. Mitchell, M. J. and Cali, A., Ultrastructural study of the development of *Vairimorpha necatrix* (Kramer, 1967) (Protozoa, Microsporida) in larvae of the corn earworm, *Heliothis zea* (Boddie) (Lepidoptera, Noctuidae) with emphasis on sporogony. *J. Eukaryotic Microbiol.*, 40, 701, 1993.

70. McLaughlin, R. E., Daum, R. J., and Bell, M. R., Development of the bait principle for boll weevil control: III. Field-cage tests with a feeding stimulant and the protozoans *Mattesia grandis* (Neogregarinida) and a microsporidian. *J. Invert. Pathol.*, 12, 168, 1968.

71. McLaughlin, R. E., Cleveland, T. C., Daum, R. J., and Bell, M. R., Development of the bait principle for boll weevil control. IV. Field tests with a bait containing a feeding stimulant and the sporozoans *Glugea gasti* and *Mattesia grandis. J. Invert. Pathol.*, 13, 429, 1969.

72. McLaughlin, R. E., Infection of the boll weevil with *Mattesia grandis* induced by a feeding stimulant. *J. Econ. Entomol.*, 59, 908, 1966.

73. McLaughlin, R. E., *Glugei gasti* sp. n., a microsporidan pathogen of the boll weevil *Anthonomous grandis. J. Protozool.*, 16, 84, 1969.

74. Stevenson, A. C. and Wenyon, C. M., Note on the occurrence of *Lankesteria culicis* in West Africa. *J. Trop. Med. Hyg.* 18, 196, 1915.

75. Samson, R. A., Vlak, J. M., and Peters, D., Fundamental and applied aspects of invertebrate pathology. *Foundation 4th Int. Colloq. Invert. Pathology*, 451, 1988.

76. Spencer, L. P. and Olson, J. K., Evaluation of the combined effects of Methoprene and the protozoan parasite *Ascogregarina culicis* (Eugregarinda, Diplocystidae) in *Aedes* mosquitoes. *Mosquito News*, 42, 384, 1982.

77. Suliaman, I., Infectivity and pathogenicity of *Ascogregarina culicis* (Eugregarinida: Lecudinidae) to *Aedes aegypti* (Diptera: Culicidae). *J. Med. Entomol.*, 29, 1, 1992.

78. Harry, O. G., Studies on the early development of the Eugregarine *Gregarina garnhami. J. Protozool.*, 12, 296, 1965.

79. Harry, O. G. and Finlayson, L. H., The life-cycle, ultrastructure and mode of feeding of the locust amoeba *Malpighamoeba locustae. Parasitology*, 72, 127, 1976.

80. Braun, L., Ewen, A. B., and Gillott, C., The life cycle and ultrastructure of *Malameba locustae* (King and Taylor) (Amoebidae) in the migratory grasshopper *Melanoplus sanguinipes* (F.) (Acrididae). *Can. Entomol.*, 120, 759, 1988.

81. Hanrahan, S. A., Ultrastructure of *Malameba locustae* (K. & T.), a protozoan parasite of locusts. *Acrida*, 4, 237, 1975.

82. Hanrahan, S. A., Further evidence for regional differentiaiton of locust *Malpighian tubule. Proc. Electron Micros. Soc. S. Afr.* 6, 31, 1976.

83. Henry, J. E., *Malameba locustae* and its antibiotoc control in grasshopper cultures. *J. Invert. Pathol.,* 11, 224, 1968.

84. King, R. L. and Taylor, A. B., *Malpighamoeba locustae,* n. sp. (Amoebidae), a protozoan parasitic in the Malpighian tubes of grasshoppers. *Trans. Am. Microsc. Soc.,* 55, 6, 1936.

85. Henry, J. E., Natural and applied control of insects by protozoa. *Annu. Rev. Entomol.,* 26, 49, 1981.

86. Davidson, E. W. and Sweeney, A. W., Review article: Microbial control of vectors: a decade of progress. *J. Med. Entomol.,* 20, 235, 1983.

87. Sajap, A. S. and Lewis, L. C., Impact of *Nosema pyrausta* (Microsporida: Nosematidae) on a predator, *Chrysoperla carnea* (Neuroptera: Chrysopidae). *Environ. Entomol.,* 18, 172, 1989.

88. Grisdale, D., An improved laboratory method for rearing large numbers of spruce budworm, *Choristoneura fumiferana* (Lepidoptera: Tortricidae). *Can. Entomol.,* 102, 1111, 1970.

89. Veber, J. and Jasic, J., Microsporidia as a factor reducing the fecundity in insects. *J. Insect Pathol.,* 3, 103, 1961.

90. Wilson, G. G., A comparison of the effects of *Nosema fumiferanae* and a *Nosema* sp. (Microsporida) on *Choristoneura fumiferanae* (Clem.) and *Choristoneura pinus pinus* (Free). *Can. For. Serv., Forest Pest Management.* Information Report, 10, 1986.

91. Sutter, G. R. and Raun, E. S., The effect of *Bacillus thuringiensis* components on the development of the European corn borer. *J. Invert Pathol.,* 8, 457, 1966.

92. Jaronski, S. T., Microsporida in cell culture. *Adv. Cell Culture* 3, 183, 1984.

93. Oien, C. T. and Ragsdale, D. W., A species specific enzyme-linked immunosorbent assay for *Nosema furnacalis* (Microspora: Nosematidae). *J. Invert. Pathol.,* 60, 84, 1992.

94. Bauer, L. S. and Nordin, G. L., Nutritional physiology of the eastern spruce budworm, *Choristoneura fumiferanae,* infected with *Nosema fumiferanae,* and interactions with dietary nitrogen. *Oecologia,* 77, 44, 1988.

10 Molecular Biology of Bacteria and Fungi for Biological Control of Weeds

Ann C. Kennedy

TABLE OF CONTENTS

I. INTRODUCTION

Weeds are a major constraint to the realization of optimum yield in agricultural systems.[1] Unwanted plant species reduce efficiency of plant establishment in range and forestry systems as well.[2,3] Synthetic chemical herbicides account for approximately 60 to 70% of all the pesticides used on cropland.[4,5] Agricultural practices,

0-8493-2442-4/96/$0.00+$.50

such as weed control with synthetic chemical herbicides, are a major cause of nonpoint source surface and ground water pollution.[4] The increased use of pesticides has intensified concern and public debate on public health and environmental consequences of current agricultural practices. Synthetic herbicides are coming under fire as contributors to problems of ground water pollution,[6] herbicide persistence,[7] and weed resistance.[8] As the public becomes more aware and concerned about pesticide residues in surface water, groundwater, and food, stricter herbicide regulations will be handed down; fewer pesticides will be released for use; and many of the pesticides presently on the market may be withdrawn. As a result, weed management strategies will be forced to rely less on synthetic chemical pesticides and more on cultural and biological methods.[9]

Biological weed control arises from the observation that biotic factors can have a significant effect on the presence and growth of plant species.[10–12] With that in mind, microorganisms that selectively suppress weed species may alter competition among plants.[13–15] These plant pathogens may be used to regulate the growth of unwanted plant species growing simultaneously with more desirable plants.[16] Biological control of weeds with microorganisms encompasses the use of fungi[13,17] and bacteria.[14] It is possible that biological control of weeds could include the use of other microbial groups, although, at this time, research specifically in this area is lacking.

It is the host/pathogen/environment interaction which determines the success of a biocontrol program. Biocontrol requires that a susceptible host weed, a favorable environment, and a virulent pathogen be present to cause weed suppression. Alteration in any one of the three components may change the disease, thus altering the biological control potential (Figure 1). The environment plays a key role by influencing the survival of the pathogen and the susceptibility of the weed to the negative effects. Genetic manipulation of the biological control agent has the potential to increase weed control. This enhancement, however, relies on continued research and investigation in this area and a greater understanding of the interactions occurring within this triumvirate.

All too often, constraints within the system limit the success of biological control programs. This lack of activity in the field can be improved with genetic technology to construct microorganisms with the desired traits and activity. There are a number of constraints in using biological control agents that need to be resolved prior to acceptance which include slow or inadequate suppression, limited host spectrum, and lack of consistency across environments. Two of the most important factors in the success of a biocontrol system is the efficacy or activity of the agent and the colonization of the host. Molecular techniques can be used to overcome these constraints by increasing virulence and competitiveness of the microorganism. Molecular techniques are also indispensable in monitoring introduced microorganisms, increasing specificity, and strengthening containment, thus reducing safety concerns. The alteration of virulence is an exciting area of research with limited published research activity at this time. The factors involved in colonization and microorganism survival are not as well defined and are difficult to identify at the molecular level. Molecular biology provides an exceptional tool for producing fungi

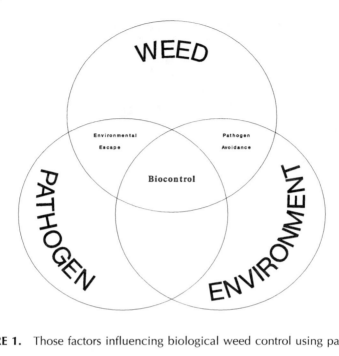

FIGURE 1.   Those factors influencing biological weed control using pathogens.

and bacteria with increased biocontrol efficacy and increasing our understanding of the mechanisms involved. Genetic technology will aid in the development of containment methods to increase agent specificity or to lessen risk of escape. Monitoring of the introduced microorganisms can also benefit from molecular investigations.

Genetic manipulation of these biocontrol agents is a tool to develop the potential to a reality, especially with those agents that would otherwise be noneffective. Genetic technology needs to be included in strategies for increasing biocontrol effectiveness. Molecular investigations of these pathogens will uncover the genetic basis of virulence and host specificity and broaden our appreciation of the range of diversity in pathogens. Genetic manipulation of the microbes can increase metabolite production, regulate host specificity or host range, and regulate production of the plant-suppressive compounds. This regulation will allow us to control the expression of plant-suppressive genes by specific promoters or under specific environmental conditions.

The use of molecular techniques in the biological control of weeds is an area that is lacking at this time, yet it is ever increasing. Strategies to improve strains for biological weed control may include investigations into pathogenicity, selectivity, competitiveness, survival, containment, and dispersal. Presently, the majority of the research being conducted is in the area of discovery of agents and development of management practices to increase biological control success. There are a number of review articles on molecular manipulations of biological control agents indicating the interest in and realization of the importance of this type of endeavor.[18–23] The challenge before us in research is to enhance these products. The purpose of this chapter is to discuss the role of molecular techniques in developing successful biocontrol strategies or improving current strategies, illustrate current research in this area, and indicate potential areas that need future research to increase efficacy and the spectrum of these valuable weed management systems.

II.   STATUS OF BIOLOGICAL WEED CONTROL

More than a hundred pathogens have been identified as having the potential for biological control of weeds.[14,24] Biological control can encompass both the classical approach and the augmentative approach. The classical approach involves the release of an agent to control weeds by its natural spread and predation. Classical strategy involves little manipulation after the initial introduction. The augmentative approach requires application of the pathogen at levels great enough to cause disease. Augmentative or inundative control methods require specific applications at certain times and doses. Suppression occurs only within the season of application and no long-term survival or control is expected. The greatest research emphasis has been in the area of fungal pathogens; however, the number of research labs involved in bacterial control measures is increasing.[14] A number of classical weed biocontrol systems have been successful, but owing to the nature of the control, commercialization was not needed. To date, three pathogens using the augmentative approach have been commercialized. Several reviews of biological control with microorganisms are available.[24–28] In order to be successful the biological control technology needs be effective, reliable, consistent, and economical.[29,30]

Biological control of weeds, in its broadest sense, has been used since the dawn of agriculture. Biological control of a weed with microbes began in the early 1970s with the introduction of a rust fungus to control rush skeleton weed.[31,32] The greatest success has been with the use of rusts (*Puccinia jaceae* Otth.) for the control of diffuse knapweed (*Centuria diffusa* Lam.),[17,33] and skeleton weed (*Chondrilla juncea*) control with the use of *Puccinia chondrillina*.[32] Other species and plant hosts are available as well.

In general, augmentative biological control agents do not reduce weeds unless applied at a high population levels. Mycoherbicides, such as the fungal pathogens sold under the trade names of Devine™, Collego™, and BioMal™ are commercially available. Devine™ is used to control stranglervine (*Morrenia odorata*) in citrus,[34] and Collego™ (*Colletotrichum gloeosporioides* f. sp. *aeschynomene*)[35] is used for the control of northern joint vetch (*Aeschynomene virginica*) in rice and soybean.[36] BioMal, the newest bioherbicide, has been registered for the control of round leaf mallow (*Malva pusilla*).[37] The commercial bioherbicides indicate the potential for the use of microorganisms in modern weed control technology.

Most of the research on microbial control of weeds has concentrated on fungal plant pathogens for broadleaf weed control. Bacteria also have a place in modern weed control, although their potential contribution often is overlooked. Soil and rhizosphere bacteria from various genera and species were found to inhibit broadleaf weeds such as velvet leaf (*Abutilon theophrasti*), morning glory (*Ipomoea* sp.), cocklebur (*Xanthium canadense* Mill.), pigweed (*Amaranthus* sp.), lambsquarter (*Chenopodium* spp.), smartweed (*Polygonum* sp.), and jimson weed (*Datura stramonium*).[38,39] The potential of biological weed control has been shown for grass weeds, as well, including downy brome (*Bromus tectorum* L.)[15] and jointed goatgrass (*Aegilops cylindrica*).[40] These bacteria were selective in laboratory, greenhouse, and field screening for their ability to inhibit weed growth while not negatively influencing the crop. Further investigation illustrated that these bacteria could be used for control of downy brome in the field[15] and are presently being explored for potential commercialization. The number of microbes and plant species that can be used in biological control programs is ever increasing.[14]

III. PATHOGENESIS

Strain development for superior weed-suppressive activity requires an understanding of the mechanisms of gene products involved in biocontrol. Molecular methods allow the identification of traits and the genes responsible for weed inhibition, thus allowing for the construction of genetically altered, superior biocontrol agents. Pathogenicity has the greatest potential for enhancement in biological weed control by molecular improvement. Progress is being made in identifying the genes and the mechanisms responsible for numerous plant pathogens.[41-43] The research in biological weed control with microorganisms presently is more involved with determining the mechanisms of action and has not yet progressed fully to the manipulation of those genes for increased activity.

A number of strategies for the improvement of biological weed control can be used to enhance pathogenicity.[44] This can include alteration in virulence, modification of specificity by increasing host range, or enhancing aggressiveness. Also, the genes responsible for detoxifying plant defenses, producing enzymes that weaken the plant cell wall, or secondary metabolite production which inhibits plant growth can be altered. Virulence can be enhanced by alteration of regulatory genes, promoter regions, or by the insertion of gene material with greater virulence into a highly competitive organism.

Several loci are thought to be involved in phytotoxin production by some pathogens, while phytotoxicity has been shown to be due to a single genetic loci in others.[45,46] Pathogenicity to a variety of plant species has been shown to be independent of one another in many cases.[46] Once the genes responsible for phytotoxic activity have been identified, insertion into the regulatory regions of another microorganism with differing host range, aggressiveness, or competitive ability is possible. Gene insertion into promoter regions can overexpress or increase copy number. *gusA* genes have been used to alter promoter regions in and enhance the expression of genes.[47] Another strategy in pathogenicity manipulation would be the addition of a gene of a known bioactivity into a competitive receptor. Insertion of the genes responsible for a nonselective toxin, such as tabtoxin, into a known pathogen may be one strategy to increase biological control activity.[48]

The genes responsible for certain enzymes, such as those involved in plant cell degradation, may be altered to increase pathogenicity. Cutinase may play a role in the aggressiveness of pathogens.[49] Insertion of the cutinase genes results in greater infection.[50] Mutants with reduced production of pectate lyase, a cell wall-degrading enzyme, caused less damage to plant cell walls than the wild type.[51] Understanding the genetics of cutinase, pectate lyase, cellulases, proteases, and other enzyme systems, will further the manipulation of these genes to enhance the pathogenicity of biological control agents. Superior strains may also be developed once mechanisms of gene products are understood.

A possible constraint to successful infection by an introduced bioherbicide may be plant defense mechanisms. Phytoalexins produced by the plant may inhibit infection and subsequent suppression of the host by biocontrol agents.[52] The genetic basis of this response and other resistance mechanisms needs further study; however, no research in this area is known at this time. This could include resistance to these compounds or possible mechanisms which reduce the antagonistic effect or suppress production of phytoalexins. The importance of phytoalexin production in biological weed control systems has been noted.[53]

IV. COMPETITIVE ABILITY/SURVIVAL

Prior to colonization and/or infection of the host by a pathogen, biocontrol agents must survive and compete in the leaf, soil, or aquatic environment with many other microorganisms. Even if a strain exhibits virulence, adaptation to the environment is necessary. Critical to the success of any biocontrol agent is its ability to withstand this initial exposure to both biotic and abiotic stresses in the soil or aerial environment for a given time period so that it may establish itself in the environment. The complexity of competitive ability and the many factors that may influence survival allow for many molecular approaches to investigations and understanding of the survival and colonization.

In order for biological control to be successful, the biological control agent needs to survive. Because of limited resources, microbes compete with each other for food, nutrients, and essential elements in the phyllosphere, soil, spermosphere, and rhizosphere.[54–56] Environmental stresses are many and include high temperature, UV radiation, moisture, osmotica, nutrients, pesticides. Soil moisture, dew period, soil texture, and temperature also have been shown to be important in influencing survival and colonization.[57–60] Those factors influencing spermosphere and root colonization are complex and many traits may be involved, therefore increasing the difficulty of study. Survival ability relies on saprophytic growth as well as colonization of host and growth on root or leaf.[60]

The biotic stresses that an introduced microorganisms will face are competition. Increased competitiveness and survival may be determined by traits such as production of antibiotics, tolerance to antibiotics, motility, resistance to starvation, or survival structures.[61,62] The ability to utilize a wide variety of substrates will increase a microorganism's resiliency. Adaptation to adverse environments may involve increased antibiotic or siderophore production which have been shown to be important in other systems.

Microbial antagonisms will influence the establishment of introduced microorganisms in the soil and on the rhizoplane, and thus are key in the biological control of many soilborne plant pathogens.[62,63,56] Antagonistic relationships demonstrated in pure culture do not always occur in the soil and the role of antibiotics in the survival and competition of microorganisms has been questioned.[63,64] Nonspecific toxin production by weed-suppressive microbes may confer a competitive advantage. Antibiotics have been suggested as increasing survival in soil,[65–67] but other substances are not thought to be important in increasing survival.[68,69] Antibiosis has been regarded as an important factor in the ecology of root-colonizing microorganisms.[70,71] In enhanced survival of biocontrol agents over a long time period, phenazine contributes to the competive of the introduced strain.[67] Investigations of phytotoxic bacteria against wheat found that the toxin did not influence competitive ability.[45]

Survival structures that widen the dispersal of the biological control agents may influence efficacy.[72] A nonsclerotial mutant of *Sclerotinia sclerotiorium* was used to study sclerotial development and survival of this fungus in the control of Canada thistle (*Cirsium arvense* scop). Lack of ascopores limits survival in adverse environmental conditions, which will reduce spread of the introduced organism, even though pathogenicity is retained.

Survival also relies on the ability of an organism to withstand stresses imposed by crop production. Resistance to benomyl, a common fungicide, may increase survival of the biological control agent *Colletotrichum gloesoporioides*, the causal agent of anthracnose, and improve the use of the mycoherbicide Collego. The gene responsible for benomyl resistance is B-tubulin (BTub) an encoding gene, *tub*1 gene. Fungicide resistant strains were obtained for tracking or determining phylogenetic relationships among *Colletotrichum* species.[61]

V.   SPECIFICITY AND CONTAINMENT

There is much discussion among researchers as to whether narrow or broad host specificity is more advantageous in the development of biological control agents. In order to minimize adverse effects to the environment, a narrow host range is recommended. However, from a marketing standpoint, a biological control agent that is too specific will not command the wide market needed to attract industry support. All too often, host-specific pathogens lack virulence and many broad host range pathogens will attack many plant species including beneficial crops. However, it may be advantageous to begin with broader host range agents to ensure efficacy, and then genetically alter the agent for reduced host range.[73] Yet in other cases, when a target host is a crop species, use of agents with broad host range would be devastating.[15,74]

Genetic technology can be used to alter host range and even increase plant production of certain toxins.[75] Several avirulence genes (*avr*) of pseudomonads and xanthomonads trigger plant resistance response and hypersensitivity and in part determine host range.[76] Also, *hrp* genes are cluster loci needed in the expression of *avr* genes and are critical to hypersensitivity and pathogenicity.[46]

Research with a broad host range pathogen illustrates the use of molecular manipulations to reduce the host range of a mycoherbicide. *Sclerotinia sclerotiorum* causes disease on both Canada thistle and spotted knapweed (*Centaurea maculosa* Lam.) Mutants with altered virulence were used to limit host range of this disease.[77] Auxotrophs, nonsclerotial forming, reduced virulence or altered host range. Pyrimidine auxotrophs have the potential to delimit the range of host or activity.[73,77] Other examples in the literature demonstrate the use of genetically altering biocontrol agents, thus increasing potential of this technique.[77–79]

In developing biological control systems, the pathogen–weed interaction is not limited and other biotic interactions can be involved as well. Narrow host range may be attained through a vector, such as an insect, which will deliver the agent to the host. In these cases host specificity may not be as critical as in other cases. The pathogen vector acts as a direct delivery system to the target host.[80,81]

The safety or ecological prudence of introducing a novel organism into an ecosystem should be of utmost importance when developing biological control agents. A driving force in the development of biological control is the additional safety and the supposed benign ecological nature of biological control agents when compared to synthetic chemicals. Biological control technology must have safety in this method of weed control. Genetic manipulation allows for increased ability to control and the ability to reduce risk that may be associated with the introduction of a foreign pathogen. One potential problem is the threat of disease or epidemics occurring on nontarget plant species, either from the introduced organism itself, the transfer of genes from the introduced to the indigenous population, or the expression of novel toxic products. Also of interest is the potential persistence or aggressiveness of introduced microorganisms which may displace the indigenous populations, thus changing the ecological balance of the ecosystem and possibly causing negative impact.

Containment of the introduced organism is possible by reducing survival and inducing auxotrophy as mentioned above.[73,82] The agent can be manipulated with regards to host range, sexual structures, metabolic activity, competitive ability, survival ability, and auxotrophy or ability to synthesize vitamins, amino acids, and aromatic fatty acids, purines, and pyrimidines.[83–88]

Several examples in the literature illustrate the potential of the use of genetic technology in altering the persistence of microorganisms and thus may be useful in containment. An auxotrophic mutant of the corn pathogen, *Cochliobolus heterostrophus*, that requires histidine for growth, does not survive as well as the wild type under field conditions.[59] Application of histidine in the field increases the population of the auxotroph and thus increases its fitness and pathogenicity. The disease in this case could be controlled by a supplemental amino acid, which controls the population of the auxotroph.

An albino mutant of that same strain exhibits virulence in the lab, but not in the field.[89] The mutant is sensitive to UV radiation and thus does not survive long enough to initiate disease. Encapsulation or protection of the mutant to reduce exposure to UV would increase its persistence, allowing it to infect the host but not allowing establishment in the ecosystem. This research indicates that virulence can be altered among plant pathogens.[43]

The presence of the T-toxin (*tox*1) gene reduced the fitness of *C. heterostrophus* when compared to an isoline not carrying the gene.[90] The insertion of such a gene, even one that may increase virulence, may in fact decrease the fitness of that microorganism allowing for manipulation for containment. The use of an auxotroph has great promise in strain improvement. Auxotrophs could be used to recombine with different traits from other auxotrophs. It is, however, important to test for virulence with or without nutrient.[77]

Compounds used as signals of exudation would have to be specific to the target. Our lack of knowledge of exudate composition may limit the use of this technique. Genes need to be identified for those processes that allow for growth on a specific carbon source, such as a soil carbon or other celluloses, xylans, pectins, and proteins. Specific inducer compounds could be identified or added to the soil for short-term toxin production or could add a microbe that either enhances or reduces substrate utilization.

Several strategies have been suggested for containment. A suicide vector which can be turned off in the presence of a specific substrate could be incorporated in the biological control agent.[91,92] Since the system is substrate driven, once the substrate is lacking, the microorganism destroys itself. For example, a suicide vector could be regulated by the presence of carbenicillin.[93] However, other substrates can function as well. A suicide vector can be associated with a gene of interest, such as the virulence gene. With such a system, coinduction of the suicidal gene with the gene of interest leading to gene expression would result in cell death. The gene product is produced, but contained. This system would be of use only when the gene product needed is limited to a point in time, and may not be useful in biocontrol where activity is required over a period of time. Multiple loci for gene expression can limit the risk involved in gene transfer and can be used for containment, which reduces the risk for not just the biocontrol product, but for the gene itself.[73] Safety is critical to the functioning of a biological control agent; however, it may not always be so alternative that it is effective yet safe. The risk involved needs to be assessed and molecular tools used to reduce this risk whenever possible.

VI. MONITORING

In order to study any organism in the environment, a tracking system is essential.[94–96] Potential artifacts need consideration before use of these techniques. Antibiotic resistance and selective media have been heavily relied upon in past ecological studies with reasonably good success, especially in monitoring of bacteria. This method assumes the microorganism can grow on selective media and the antibiotic resistant trait is stable and does not influence other microbial characteristics. Also, the issue of viable, but not culturable, populations arises with these monitoring techniques.[97,98] Detection levels, unfortunately, may be quite high, thus reducing the power of culturable methods. These potential problems in the classical methods of monitoring make molecular techniques for monitoring weed-suppressive microorganisms even more appealing. Many types of molecular markers can be identified and used in the detection of an individual and to identify genes of interest. There are a number of markers or tracers available with specific genetic traits unique to the system that give visual or measurable signals. Nucleic acid probes and phylogenetic determinations also may be helpful in monitoring weed-suppressive microorganisms.

A.   GENETIC TRAITS

A *lac* Z and *lac* Y marker system was developed by the Monsanto Company to monitor the environmental release of pseudomonads.[99,100] The lac operons of *E. coli*, *lac* Z and *lac* Y were inserted into *P. aureofaciens*. Lactose utilization is a trait not normally present in *P. aureofaciens* or in soil microorganisms. The lac ZY gene product cleaves the chromogenic dye x-gal yielding blue color to colonies on the specific isolation medium.[99] Field trials confirmed the effectiveness of detection and limited carryover. The trials also established that no detectable transfers of the lac ZY gene occurred to indigenous flora.[101] The TN7::lac ZY insertion, not present in the soil, can be used to probe for DNA by soil extraction and DNA hybridization.[99-102]

The ice nucleating gene, *ina Z*, has been used to identify introduced species.[103] Insertion of this gene allows for the enumeration of soil bacteria and *in situ* expression without having to culture.[104] Ice-nucleating activity can be quantitative, correlated with the amount of inaZ protein present. Soil and aquatic bacteria do not produce this activity, thus those with the gene insert can be enumerated and gene expression followed.

Another means of tracking microbes is the lux reporter system which codes for enzymes responsible for the light-emitting reaction, bioluminescence.[105-107] The photons emitted by the bacteria expressing this gene can be counted by a scintillation counter or lumitor. In theory, a single cell can be detected through bioluminescence by charged coupling device enhanced microscopy.[108]

B.   GENETIC PROBES AND ANALYSES

Gene analysis represents a powerful tool for monitoring and identifying microorganisms in the environment. Nucleic acid hybridization has been successfully used in the detection of bacteria and fungi.[109-111] Nucleic acid-based analyses have expanded our knowledge of plant pathogenic characteristics and further refined our monitoring of these microorganisms.

Restriction fragment length polymorphism (RFLP) can be used to assess the genetic variability and taxonomic relationships of plant pathogens.[112-114] RFLP can also provide markers for detection; however, DNA preparation is time consuming and difficult, and faster, and easier methods are continually appearing in the literature. Although RFLP may be limited as a diagnostic tool, it is a valuable technique in assessing populations and relationships among weed-suppressive microorganisms.

Polymerase chain reaction (PCR) amplification of nucleic acids is a highly sensitive technique for use in detection of unique sequences from environmental samples.[115–118] PCR increases the sensitivity of DNA probes and allows for DNA or RNA detection at a lower copy number. PCR is a technique that can be used in biological weed control monitoring and in research on taxonomic and phylogenetic relationships among this group of microorganisms.

Phylogenetic analysis of weed-suppressive microbes may be an important tool in monitoring and in developing a greater understanding of these microorganisms. Phylogenies developed using gene sequences have proven to be of great utility in classifying and identifying microorganisms.[119–123] When a trait or microbial group is defined phylogenetically, oligonucleotides targeted against portions of a gene can be used to screen soil, plant, or other environmental samples for the specific microorganism and closely related microorganisms. Diagnostic probes will be useful in monitoring the effects of environmental conditions and treatments on introduced strains and related microbes in the field. Also, phylogenies will lead to development of group- and/or strain-specific diagnostic probes for laboratory screening for related strains that may have similar weed-suppressive properties. Group- and species-specific probes have been used in successful diagnostic tests for a wide range of fungi and bacteria.[111,120–124] An extensive database of specific DNA and RNA sequences has been developed along with rapid methods for isolation, sequencing, and characterization.[111,122] Phylogenetic evaluations may identify close relatives that have been well studied, which may provide insights into the mechanisms of suppression and survival. The versatility of this technique can be seen for specific genus identification using the conserved region and species specific identification using the nonconserved region.[122,125,126]

Molecular investigations into monitoring weed-suppressive microorganisms allow researchers to identify the traits involved and determine their status within the soil community. These probes will also assist researchers in monitoring the effects of environmental perturbations and management on specific microbes as well as the total population of weed-suppressive microorganisms. Molecular techniques expand our knowledge base to include those soil inhabitants that are nonculturable. The research to date in the area of monitoring indicates the usefulness, yet limitations, of all techniques in determining the mechanisms of biological weed control.

VII. CONCLUSION

Genetic manipulation of biological weed control agents will lead to greater use of these agents to reduce weed populations and reduce our reliance on synthetic chemical herbicides. Many challenges lie ahead for those involved in the research on microorganisms for weed control. No longer can we function only as collectors of species with suspected biological control activity. Future research endeavors will need to include many areas, none as important as pathogenicity and competitive ability. Molecular manipulations can be helpful in determining the mechanisms of action of these phytotoxins and increasing our understanding of the functioning of these microorganisms to further enhance the activity or specificity of these plant-inhibitory compounds. A deeper understanding of the ecology of the introduced bacteria can be achieved through molecular manipulations. This research is imperative to increase survival and efficacy, and to improve the effectiveness of biological control agents. A greater understanding of the ecological constraints of each system will also foster development of delivery systems or formulations that will reduce the impact of biotic and abiotic forces and increase the viability, efficacy, and ease of application. Microbial detection methods, at present, are inadequate, and enumeration methods and monitoring can be enhanced by molecular probes and analyses.

The number of microorganisms, both fungi and bacteria, which have potential for biological weed control is endless. Genetic manipulation of these biological control agents will allow us to alter the ecological balance to favor the pathogen and suppression of the weed. This area of research needs greater funding and research activity. Molecular investigations are imperative in the development of microorganisms for biological control to reduce synthetic chemical pesticide use.

ACKNOWLEDGMENTS

Support from the U.S.D.A.–Agricultural Research Service in cooperation with the College of Agriculture and Home Economics, Washington State University, Pullman, WA 99164 is greatly appreciated. Trade names and company names are included for the benefit of the reader and do not imply endorsement or preferential treatment of the product by the U.S. Department of Agriculture. All programs and services of the U.S. Department of Agriculture are offered on a nondiscriminatory basis without regard to race, color, national origin, religion, sex, age, marital status, or handicap.

REFERENCES

1. U.S. Department of Agriculture, Economic Indicators of the Farm Sector: Costs of Production, ECIFS 6-1. Washington, D.C., 1988, 128.
2. Lacey, C. A., Lacey, J. R., Chicoine, T. K., Fay, P. K., and French, R. A., Controlling knapweed on Montana rangeland, *Mont. State Univ., Coop Ext. Serv. Circ.*, No. 311, 1986.
3. Chan, S. S. and Walsted, J. D., Correlations between overtopping vegetation and development of douglas fir in the Oregon Coast Range West, *J. Appl. For.*, 2, 117, 1987.
4. National Research Council, Alternative Agriculture, National Academic Press, Washington, D.C., 1989, 448.
5. Gianessi, L. P. and Puffer, C., *Herbicide use in the United States: National Summary Report*, Resources for the Future, Washington D. C., 1991.
6. Carter, A. D., Hollis, J. M., Thompson, T. R., Oakes, D. B., and Binney, R., Pesticide contamination of water sources: current policies for protection and a multidisciplinary proprosal to aid future planning, in *Brighton Crop Protection Conference-Weeds*, England, Nov. 1991, Vol. 2, 491, 1991.

7. Stougard, R. N., Shea, P. J., and Martin, A. R., Effect of soil type and pH on adsorption, mobilitiy and efficacy of imazaquin and imazethapyr, *Weed Sci.*, 38, 67, 1990.

8. Holt, J. S. and LeBaron, H. M., Significance and distribution of herbicide resistance, *Weed Technol.*, 4, 141, 1990.

9. Youngberg, G. and Ridgway, R., Biologically based methods of pest control: Contributions to a sustainable agriculture, *Am. J. Alternative Agric.*, 3, 50, 1988.

10. Ohr, H. D., Plant disease impacts on weeds in the natural ecosystem, *Proc. Am. Phytopathol. Soc.*, 1, 181, 1974.

11. Dinoor, A. and Eshed, N., The role and importance of pathogens in natural plant communities, *Annu. Rev. Phytopathol.*, 22, 443, 1984.

12. Quimby, P. C., Jr., Impact of diseases on plant populations, in *Biological Control of Weeds with Plant Pathogens*, Charudattan, R. and Walker, H. L., Eds., Wiley Press, New York, 1982, 47.

13. Hasan, S. and Ayers, P. G., The control of weeds through fungi: principles and prospects, *New Phytol.*, 115, 201, 1990.

14. Kremer, R. J. and Kennedy, A. C., Rhizobacteria as biocontrol agents of weeds, *Weed Technol.*, in press.

15. Kennedy, A. C., Elliott, L. F., Young, F. L., and Douglas, C. L., Rhizobacteria suppressive to the weed downy brome, *Soil Sci. Soc. Amer. J.*, 55, 722, 1991.

16. Templeton, G. E., Status of weed control with plant pathogens, in *Biological Control of Weeds with Plant Pathogens*, Charudattan, R. and Walker, H. L., Eds., Wiley Press, New York, 1982, 29.

17. Mortensen, K., Biological control of weeds with plant pathogens, *Can. J. Plant Pathol.*, 8, 229, 1986.

18. Betz, F. S., Genetically engineered microbial pesticides, in *Biotechnology for Crop Protection*, Hedin, P. A., Menn, J. J., and Hollingworth, R. M., Eds., ACS Symposium Series - Am. Chem. Society, The American Chemical Society, Washington, D.C., 1988, 437.

19. Charudattan, R., The use of natural and genetically altered strains of pathogens for control, in *Biological Control in Agricultural IPM Systems*, Hoy, M. A. and Herzog, D. C., Eds., Academic Press, New York, 1985, 347.

20. Kistler, C. H., Genetic manipulation of plant pathogenic fungi, in *Microbial Control of Weeds*, TeBeest, D. O., Ed., Chapman and Hall, New York, 1991, 152.

21. Lacy, G. H., Perspectives for biological engineering of prokaryotes for biological control of weeds, in *Microbial Control of Weeds*, TeBeest, D. O., Ed., Chapman and Hall, New York, 1991, 135.

22. Perlak, F. J., Obukowicz, M. G., Watrud, L. S., and Kaufman, R. J., Development of genetically engineered microbial biocontrol agents, in *Biotechnology for Crop Protection*, Hedin, P. A., Menn, J. J., and Hollingworth, R. M., Eds., ACS Symposium Series - Am. Chem. Society, The American Chemical Society, Washington, D.C., 1988, 284.

23. Sands, D. C., Miller, R. V., and Ford, E. J., Biotechnological approaches to control of weeds with pathogens, in *Microbes and Microbial Products as Herbicides*, Hoagland, R. E., Ed., American Chemical Society, Washington, D.C., 1990, 184.

24. Charudattan, R., The mycoherbicide approach with plant pathogens, in *Microbial Control of Weeds*, TeBeest, D. O., Ed., Chapman and Hall, London, 1991, 24.

25. Harris, P., Classical biocontrol of weeds, its definition, selection effective agents, and administrative-political problems, *Can. Entomol.*, 123, 827, 1991.

26. TeBeest, D. O., Yang, X. B., and Cisar, C. R., The status of biological control of weeds with fungal pathogens, *Annu. Rev. Phytopathol.*, 30, 637, 1992.

27. TeBeest, D. O., *Microbial Control of Weeds*, Chapman and Hall, New York, 1991.
28. Templeton, G. E. and Trujillo, E. E., The use of plant pathogens in the biological control of weeds, in *CRC Handbook of Pest Management In Agriculture*, Pimentel, D., Ed., CRC Press, Boca Raton, FL, 1981, 345.
29. Hasan, S., Industrial potential of plant pathogens as biocontrol agents of weeds, *Symbiosis*, 2, 151, 1986.
30. Harman, G. E., Biocontrol of weeds with microbes, in *Biological Control of Plant Disease*, Mukerji, K. G. and Garg, K. L., Eds., CRC Press, Boca Raton, FL, 1988, 129.
31. Adams, E. B. and Line, R. F., Biology of *Puccinia chondrillina* in Washington, *Phytopathology*, 74, 742, 1984.
32. Cullen, J. M., Kable, P. F., and Catt, M., Epidemic spread of a rust imported for biological control, *Nature*, 244, 462, 1973.
33. Watson, A. K. and Clement, M., Evaluation of rust fungi as biological control agents of weedy *Centaurea* in North America, *Weed Sci.*, 34, 7, 1986.
34. Ridings, W. H., Biological control of stranglervine (*Morrenia odorata* Lindl.) in citrus — a researcher's view, *Weed Sci.*, 34, Supp. 1, 1986.
35. Daniel, J. T., Templeton, G. E., Smith, R. J., Jr., and Fox, W. T., Biological control of northern jointvetch in rice with an endemic fungal disease, *Weed Sci.*, 21, 303, 1973.
36. Templeton, G. E., TeBeest, D. O., and Smith, R. J., Biological weed control with mycoherbicides, *Annu. Rev. Phytopathol.*, 17, 301, 1979.
37. Grant, N. T., Prusinkiewicz, E., Mortensen, K., and Makowski, R. M. D., Herbicide interactions with *Colletotrichum gloeosporioides* f. sp. *malvae*, a bioherbicide for round-leaved mallow (*Malva pusilla*) control, *Weed Technol.*, 4, 716, 1990.
38. Kremer, R. J., Antimicrobial activity of velvetleaf (*Abutilon theophrasti*) seeds, *Weed Sci.*, 34, 617, 1986.
39. Kremer, R. J., Begonia, M. F. T., Stanley, L., and Lanham, E. T., Characterization of rhizobacteria associated with weed seedlings, *Appl. Environ. Microbiol.*, 56, 1649, 1990.
40. Kennedy, A. C., Ogg, A. G., Jr., and Young, F. L., U.S. Patent 07/597,150, Biocontrol of Jointed Goatgrass, November 17, 1992.
41. Bronson, C. R., The genetics of phytotoxin production by plant pathogenic fungi, *Experientia*, 47, 771, 1991.
42. Willis, D. K., Barta, T. M., and Kinscherf, T. G., Genetics of toxin production and resistance in phytopathogenic bacteria, *Experientia*, 47, 766, 1991.
43. Yoder, O. C., Altered virulence in recombinant fungal pathogens, *Proc. 5th Int. Congr., Plant Pathol.*, Kyoto, Japan, August 18–27, 1988, 226.
44. Roberts, D. P. and Fravel, D. R., Strategies and techniques for improving biocontrol soilborne plant pathogens, *ACS Symposium Series*, Washington, D.C., 1993, 323.
45. Kennedy, A. C., Bolton, H. Jr., Stroo, H. F., and Elliott, L. F., The competitive abilities of the Tn5 Tox– mutants of a rhizobacterium inhibitory to wheat growth, *Plant Soil*, 144, 143, 1992.
46. Lindgren, P. B., Panopoulos, N. J., Staskawicz, B. J., and Dahlbeck, D., Genes required for pathogenicity and hypersensitivity are conserved and interchangeable among pathovars of *Pseudomonas syringae*, *Mol. Gen. Genet.*, 211, 499, 1988.
47. Broglie, B. and Roby, C.H., Model suicide vector for containment of genetically engineered microorganisms, *Appl. Environ. Microbiol.*, 54, 2472, 1988.
48. Brooker, N. L., Mischke, C. F., Mischke, S., and Lydon, J., Transformation of *Colletotrichum* with the acetylation gene (bar) from Streptomyces, *Fungal Genet. Newsl.*, 40A, 36, 1993.

49. Dickman, M. B., Patil, S. S., and Kolattukudy, P. E., Purification, characterization and role in infection of an extracellular cutinolytic enzyme from *Colletotrichum gloeosporioides* Penz, on *Carica papaya* L., *Physiol. Plant Pathol.*, 20, 333, 1982.

50. Ettinger, W. F., Thukral, S. K., and Kolattukudy, P. E., Structure of cutinase gene, cDNA, and the derived amino acid sequence from phytopathogenic fungi, *Biochemistry* 26, 7883, 1987.

51. Reid, J. L., and Collmer, A., Construction and characterization of an *Erwinia chrysanthemi* mutant with directed deletions in all of the pectate lyase structural genes, *Mol. Plant-Microbe Interact.*, 1, 32, 1988.

52. Van Etten H. D., Matthews, D. E., and Mackintosh, S. F., Adaptation of pathogenic fungi to toxic chemical barriers in plants: the psiatin demethylase of *Nectria haematoccocca* as an example, in *Molecular Strategies for Crop Protection*, UCLA Symposium on Molecular and Cellular Biology, A. R. Liss, New York, 1987, 48, 59.

53. Sharon A. and Gressel, J., Elicitation of a flavonoid phytoalexin accumulation in *Cassia obtusifolia* by a mycoherbicide: estimation by $AlCl_3$- spectrofluorimetry, *Pestic. Biochem. Physiol.*, 41, 1442, 1991.

54. Baker, R., Eradication of plant pathogens by adding organic amendments to soil, in *Handbook of Pest Management in Agriculture*, Pimental, D., Ed., CRC Press, Boca Raton, FL, 1981, 137.

55. Garrett, S. D., *Pathogenic Root-Infecting Fungi*, Cambridge University Press, London, 1970.

56. Weller, D. M., Biological control of soilborne plant pathogens in the rhizosphere with bacteria, *Annu. Rev. Phytopathol.*, 26, 379, 1986.

57. Elad, Y. and Chet, I., Possible role of competition for nutrients in biocontrol of *Pythium* damping-off by bacteria, *Phytopathology*, 77, 190, 1987.

58. Johnson, B. N., Kennedy, A. C., and Ogg, A. G., Jr., Suppression of downy brome growth by a rhizobacterium in controlled environments, *Soil Sci. Soc. Am. J.*, 57, 73, 1993.

59. Garber, R. C., Fry, W. E., and Yoder, O. C., Conditional field epidemics on plants: a resource for research in population biology, *Ecology*, 64, 1653, 1983.

60. Ahmad, J. S. and Baker, R., Competitive saprophytic ability and cellulolytic activity of rhizosphere-competent mutants of *Trichoderma harzianum*, *Phytopathology*, 77, 358, 1987.

61. Panaccione, D. G., McKiernan, M., and Hanau, R. M., *Colletotrichum graminicola* transformed with homologous and heterologous benomyl-resistant genes retains expected pathogenicity to corn, *Mol. Plant-Microbe Interact.*, 1, 113, 1988.

62. Fravel, D. R., Role of antibiosis in the biocontrol of plant diseases, *Annu. Rev. Phytopathol.*, 26, 75, 1988.

63. Thomashow, L. S., Weller, D. M., Bonsall, R. F., and Pierson, L. S., Production of the antibiotic phenazine-1-carboxylic acid by fluorescent *Pseudomonas* species in the rhizosphere of wheat, *Appl. Environ. Microbiol.*, 56, 908, 1990.

64. Williams, S. T. and Vickers, J. C., The ecology of antibiotic production, *Microb. Ecol.*, 12, 43, 1986.

65. Atlas, T. M. and Bartha, R., *Microbial Ecology: Fundamentals and Applications*, Benjamin-Cummings, Menlo Park, CA, 1987.

66. Bruehl, G. W., Millar, R. L., and Cunfer, B., Significance of antibiotic production by *Cephalosporium gramineum* to its saprophytic survival, *Can. J. Plant Sci.*, 49, 235, 1969.

67. Mazzola, M., Cook, J. R., Thomashow, L. S., Weller, D. M., and Pierson, L. S., Contribution of phenazine antibiotic biosynthesis to the ecological competence of fluorescent pseudomonads in soil habitats, *Appl. Environ. Microbiol.*, 58, 8, 1992.

68. Williams, S. T., Are antibiotics produced in soil, *Pedobiologia*, 23, 427, 1982.
69. Gottlieb, D., The production and role of antibiotics in soil, *J. Antibiot.*, 29, 987, 1976.
70. Brian, P. W., Effects of antibiotics on plants, *Annu. Rev. Plant. Physiol.*, 8, 413, 1957.
71. Fravel, D. R. and Lewis, J. A., Production, formulation and delivery of beneficial microbes for biocontrol of plant pathogens, *Pestic. Formulations Appl. Syst.*, 11, 173, 1992.
72. Miller, R. V., Ford, E. J., Zidack, N. K., and Sands, D. C., A pyrimidine auxotroph of *Sclerotinia sclerotiorum* for use in biological weed control, *J. Gen. Microbiol.*, 135, 2085, 1989.
73. Sands, D. C., Ford, E. J., and Miller, R. V., Genetic manipulation of broad host-range fungi for biological control of weeds, *Weed Technol.*, 4, 471, 1990.
74. Fredrickson, J. K. and Elliott, L. F., Colonization of winter wheat roots by inhibitory rhizobacteria, *Soil Sci. Soc. Amer. J.*, 49, 1172, 1985a.
75. Riddle, D. L. and Georgi, L. L., Advances in research on *Caenorhabditis elegans*: application to plant parasitic nematodes, *Annu. Rev. Phytopathol.*, 28, 247,1990.
76. Staskawicz, B. J., Dahlbeck, D., Keen N., and Napoli, C., Molecular characterization of cloned avirulence gene from race 0 and race 1 of *Pseudomonas syringae* pv *glycineae*, *J. Bacteriol.*, 169, 5789, 1987.
77. Miller, R. V., Ford, E. J., and Sands, D. C., A nonsclerotial pathogenic mutant of *Sclerotinia sclerotiorum*, *Can. J. Microbiol.*, 35, 517, 1989.
78. Anderson, D. M. and Mills, D., The use of transposon mutagenesis in the isolation of nutritional and virulence mutants in two pathovars of *Pseudomonas syringae, Phytopathology*, 75, 104, 1985.
79. Tinline, R. D., *Cochliobolus sativus* VII. Nutritional control of the pathogenicity of some auxotrophs to wheat seedlings. *Can. J. Bot.*, 41, 489, 1963.
80. Kremer, R. J. and Spencer, N. R., Impact of a seed-feeding insect and microorganisms on velvetleaf *Abutilon theophrasti* seed viability, *Weed Sci.*, 37, 211, 1989.
81. Purcell, A. H., Evolution of the insect vector relationship in *Phytopathogenic Prokaryotes*, Vol. 1, Mount, M. S. and Lacy, G. H., Eds., Academic Press, New York, 1982, 121.
82. Knudsen, S. M. and Karlstrom, O. H., Development of efficient suicide mechanisms for biological containment of bacteria, *Appl. Environ. Microbiol.*, 57, (1), 85, 1991.
83. Colbert, S. F., Isakeit, T., Ferri, M., Weinhold, A. R., Hendson, M., and Schroth, M. N., Use of an exotic carbon source to selectively increase metabolic activity and growth of *Pseudomonas putida* in soil, *Appl. Environ. Microbiol.*, 59, 7, 1993.
84. Contreras, A., Molin, S., and Ramos, J., Conditional suicide containment system for bacteria which mineralize aromatics, *Appl. Environ. Microbiol.*, 57, 1504, 1991.
85. Jensen, L. B., Ramos, J. L., Kaneva, Z., and Molin, S., A substrate dependent biological containment system for *Pseudomonas putida* based on the *Escherichia coli gef* gene, *Appl. Environ. Microbiol.*, 59, 3713, 1993.
86. Recorbet, G., Robert, C., Givaudan, A., Kudlab, B., Normand, P., and Faurie, G., Conditional suicide system of *Escherichia coli* released soil that uses the *Bacillussubtilis* sacB gene, *Appl. Environ. Microbiol.*, 59, 1361, 1993.
87. Slininger, P. J. and Jackson, M. A., Nutritional factors regulating growth and accumulation of phenazine 1-carboxylic acid by *Pseudomonas fluorescens* 2-79, *Appl. Microbiol. Biotechnol.*, 37, 388, 1992.
88. Schisler, D. A., Jackson, M. A., and Bodnost, R. J., Influence of nutrition during conidiation of *Colletotrichum truncatum* on conidial germination and efficacy in inciting disease in *Sesbania exaltata*, *Phytopathology Biotechnol.*, 37, 388, 1992.
89. Yoder, O. C., *Cochliobolus heterostrophus* cause of southern corn leaf blight, in *Genetics of Plant Pathogenic Fungi, Advances in Plant Pathology*, Vol. 6, Academic Press, San Diego, CA, 1988, 93.

90. Klittich, C. J. R. and Bronson, C. R., Reduced fitness associated with *tox*1 of *Cochliobolus heterostrophus, Phytopathology*, 76, 1294, 1986.
91. Thomson, J. A., Biological control of plant pests and pathogens: alternative approaches, in *Biotechnology in Plant Disease Control*, Chet, I., Ed., Wiley-Liss, New York, 1993, 275.
92. McCormick, D., The circle of our felicity, *Biotechnology*, 4, 751, 1986.
93. Bej, A. K., Perlin, M. H., and Atlas, R. M., Model suicide vector for containment of genetically engineered microorganisms, *Appl. Environ. Microbiol.*, 54, 2472, 1988.
94. Hwang, I. and Farrand, S. K., A novel gene tag for identifying microorganisms released into the environment, *Appl. Environ. Microbiol.*, 60, 913, 1994.
95. Kluepfel, D. A., The behavior and tracking of bacteria in the rhizosphere, *Annu. Rev. Phytopathol.*, 31, 441, 1993.
96. Michelmore, R. W. and Hulbert, S. H., Molecular markers for genetic analysis of phytopathogenic fungi, *Annu. Rev. Phytopathol.*, 25, 383, 1987.
97. Colwell, R. R., Brayton, P. R., Grimes, D. J., Roszak, D. B., and Huq, S. A., Viable but not culturable *Vibrio cholerae* and related pathogens in the environment, implications for release of genetically engineered microorganisms, *Bio/Technology*, 3, 817, 1985.
98. Roszak, D. B. and Colwell, R. R., Survival strategies of bacteria in the natural environment, *Microbiol. Rev.*, 51, 365, 1987.
99. Barry, G. F., Permanent insertion of foreign genes into the chromosomes of soil bacteria, *Bio/Technology*, 4, 446, 1986.
100. Drahos, D. J., Barry, G. F., Hemming, B.C., Brandt, E. J., Kline, E. L., Skipper, H. D., and Kluepfel, D. A., Spread and survival of genetically marked bacteria in the soil, in *Release of Genetically Engineered and Other Microorganisms*, Fry, J. C. and Day, M. J., Eds., Cambridge University Press, New York, 1992, 178.
101. Kluepfel, D. A., Kline, E. L., Skipper, H. D., Hughes, T. A., and Gooden, D. T., Release and tracking of genetically engineered bacteria in the environment, *Phytopathology*, 81, 328, 1991.
102. Hofte, M., Mergeay, M., and Verstraete, W., Marking the rhizopseudomonas strain 7NSK₂ with a Mu d(*lac*) element for ecological studies, *Appl. Environ. Microbiol.*, 56, 1046, 1990.
103. Loper, J. E., Henckels, M. D., and Lindow, S. E., A biological sensor for iron that is available to *Pseudomonas fluorescens* inhabiting the plant rhizosphere, in *Advances in Molecular Genetics of Plant-Microbe Interactions*, Vol. 2, Nester, E. W. and Verma, D. P. S., Eds., Kluwer Academic Publishers, Dordrecht, 1993, 543.
104. Lindow, S. E., Amy, D. C., and Upper, C. D., *Erwinia herbicola*: an active ice nucleus incites frost damage to maize, *Phytopathology*, 68, 523, 1978.
105. de Weger, L. A., Dunbar, P., Mahafee, W. F., Lugtenburg, B. J., and Sayler, G. S., Use of bioluminescence markers to detect *Pseudomonas* bacteria in the rhizosphere, *J. Appl. Environ. Microbiol.*, 57, 3641, 1991.
106. King, J. M. H., Digrazia, P. M., Applegate, B., Burlage, R., Sanseverino, J., Dunbar, P., Larimer, F., and Sayler, G. S., Rapid sensitive bioluminescent reporter technology for naphthalene exposure and biodegradation, *Science*, 249, 778, 1990.
107. Stewart, G. S. and Williams, A. B. P., *lux* genes and the applications of bacterial luminescence, *J. Gen. Microbiol.*, 138, 1289, 1992.
108. Silcock, D. J., Waterhouse, R. N., Glover, L. A., Prosser, J. I., and Kilham, K., Detection of a single genetically modified bacterial cell in soil by using charge coupled device-enhanced microscopy, *Environ. Microbiol.*, 58, 2444, 1992.
109. Atlas, R. M., Sayler, G., Burlage, R. S., and Bej, A. K., Molecular approaches for environmental monitoring of microorganisms, *Biotechniques*, 12, 706, 1992.

110. Liesack, W. and Stackebrandt, E., Occurrence of novel groups of the domain bacteria as revealed by analysis of genetic material isolated from an Australian terrestrial environment, *J. Bacteriol.*, 174, 5072, 1992.

111. Ward, D., Bateson, M., Weller, R., and Ruff-Roberts, A., Ribosomal RNA analysis of microorganisms as they occur in nature, *Adv. Microb. Ecol.*, 12, 219, 1992.

112. Graham, J. H., Hartung, J. S., Stall, R. E., and Chase, A. R., Pathological, restriction fragment length polymorphisms and fatty acid profile relationships between *Xanthomonas campestris* from citrus and noncitrus hosts, *Phytopathology*, 80, 829, 1990.

113. Koch, E., Song, K., Osborn, T. C., and Williams, P. H., Relationships between pathogenicity and phylogeny based on restriction fragment length polymorphisms in *Leptosphaeria maculans, Mol. Plant-Microbe Interact.*, 4, 341, 1991.

114. MacDonald, B. A. and Martinez, J. P., DNA restriction fragment polymorphisms among *Mycosphaerella graminicola* (anamorph *Septoria tritici*) isolates collected from a single wheat field *Phytopathology*, 74, 655, 1990.

115. Bruce, K. D., Hiorns, W. D., Hobman, J. L., Osborn, A., and Strike, A. M., Amplification of DNA from native populations of soil bacteria by using the polymerase chain reaction, *Appl. Environ. Microbiol.*, 58, 3413, 1992.

116. Picard, C., Ponsonnet, C., Paget, E., Nesme, X., and Simonet, P., Detection and enumeration of bacteria in soil by direct DNA extraction and polymerase chain reaction, *Appl. Environ. Microbiol.*, 2717, 1992.

117. Rollo, F., Salvi, R., and Torchia, P., Highly sensitive and fast detection of *Phoma tracheiphila* by polymerase chain reaction, *Appl. Microbiol. Biotechnol.*, 32, 572, 1990.

118. Steffan, R. and Atlas, R. M., DNA amplification to enhance detection of genetically engineered bacteria in environmental samples, *Appl. Environ. Microbiol.*, 54, 2185, 1988.

119. DeLong, E. F., Wickham, G. S., and Pace, N. M., Phylogenic strains: ribosomal RNA based probes for the identificaiton of single cells, *Science*, 243, 1360, 1989.

120. Giovannoni, S., De Long, E., Olsen, G., and Pace, N., Phylogenetic group specific probes for identification of single microbial cells, *J. Bacteriol.*, 170, 720, 1988.

121. Olsen, G. J., Lane., D. J., Giovannoni, S. J., Pace, N. R., and Stahl, D. A., Microbial ecology and evolution: a ribosomal RNA approach, *Annu. Rev. Microbiol.*, 40, 337, 1986.

122. Stahl, D. A., Flesher, B., Mansfield, H. R., and Montgomery, L., Use of phylogenetically based hybridization probes for studies of ruminal microbial ecology, *Appl. Environ. Microbiol.*, 54, 1079, 1988.

123. Woese, C. R., Stackebrandt, E., Macke, T. J., and Fox, G. E., A major phylogenetic definition of the major eubacterial taxa, *Syst. Appl. Microbiol.*, 6, 251, 1985.

124. Hahn, D., Amann, R. I., Ludwig, W., Akkermans, A. D. L., and Schleifer, K. H., Detection of micro-organisms in soil after *in situ* hybridization with rRNA-targeted, fluorescently labelled oligonucleotides, *J. Gen. Microbiol.*, 138, 879, 1992.

125. Rasmussen, O. F. and Reeves, J., DNA probes for the detection of plant pathogenic bacteria, *J. Biotech.*, 25, 203, 1992.

126. Voordouw, G., Voordouw, J. K., Karkhoff-Schweizer, R. R., Fedorak, P. M., and Westlake, D. M. S., Reverse sampling genome probing, a new technique for identification of bacteria in environmental samples by sDNA hybridization, and its application to the identification of sulfate-reducing bacteria in oil field samples, *Appl. Environ. Microbiol.*, 57, 3070, 1991.

11 Integrated Pest Management Strategies

V. Ragunathan and B. J. Divakar

TABLE OF CONTENTS

0-8493-2442-4/96/$0.00+$.50
© 1996 by CRC Press, Inc.

I. INTRODUCTION

Asian agriculture has gone through tremendous change during the later half of the 20th century, leading to the "green revolution" in foodgrains. Yet there are hundreds of millions of people in many parts of the world who suffer from acute shortage of food and from malnutrition. Poor productivity and lack of knowledge about various aspects of agricultural inputs and crop management are the major reasons for this food shortage. In tropical countries, pests proliferate rapidly and sometimes inflict heavy loss to crops in the fields as well as in storage. The spectacular results obtained in crop protection with the use of BHC and DDT during 1947–1948 led to the discovery and synthesis of several other molecules, and the farming community gradually became more and more dependent on synthetic pesticides. The extensive and excessive use of these pesticides also led to several problems, such as destruction of beneficial organisms, development of resistance in some species, resurgence of some pests, environmental pollution, and health hazards due to pesticide residues in the food chain. In the State of Andhra Pradesh in India, continuous failure of the cotton crop during 1985 through 1987 due to whitefly and bollworm *(Helicoverpa armigera),* which have developed multifold resistance to most of the insecticides used, led to the sad suicide of scores of farmers, unable to face the economic crisis. Thousands of innocent lives were lost and some of the affected people are still suffering, due to an industrial hazard that occurred in a pesticide industry at Bhopal in India during 1984. There are many such instances of accidental poisoning all over the world, most of which go unrecorded. The chronic poisoning due to repeated or long-term exposure to small amounts of toxicant is scarcely noticed. About half the poisoning cases and nearly three quarters of the deaths are estimated to occur in developing countries. The International Organisation of Consumer Unions put the figure for 1986 at 375,000 human poisoning cases in developing countries, of whom 10,000 died. About 20% of all farmers in developing countries suffer from pesticide intoxication at least once in their working life.

Considering the adverse effects of over-reliance on pesticides in many developing countries, more emphasis is now given to popularise Integrated Pest Management for tackling pest problems in crops like cotton, rice, vegetables, etc.

Integrated pest management (IPM) has been defined as "A pest management system which in the context of the associated environment and population dynamics of the pest species, utilises all suitable techniques in as compatible a manner as possible to maintain the pest population at levels below those causing economic injury (cf. Oudejans, 1991)." It is an economically justified system of crop protection which combines several techniques such as cultural, mechanical, biological, and chemical control methods to sustain productivity with minimum adverse effect on the environment. Whenever chemical control is needed, careful selection and restrained use of pesticides is essential to avoid harmful interaction. However, preference should be given to other methods of control first.

II.   CULTURAL CONTROL

A.   AGRONOMIC PRACTICES

Many agronomic practices can reduce pest populations to a great extent. Such practices were the only tools of pest management available to the farmer before the advent of synthetic pesticides. Therefore, in the IPM strategy, adopting cultural practices is the first step in crop protection. Repeated ploughing and deep ploughing expose many insect pests which spend part of their life cycle in the soil. Whitegrubs, cutworms, and pupal stages of lepidopterans are thus exposed to adverse climatic conditions and also to predation by insectivorous birds.

B.   CROP ROTATION AND INTERCROPPING

Cultural practices modify the cropping environment in such a way that it becomes less favourable for the development of the pest population. A suitable crop rotation will interrupt the relationship between the pest and the host plant. Summer cultivation of maize or sorghum has been found to be useful in minimising the damage in cotton by *Spodoptera litura*. Cyst nematode of potato, *Globodera* sp., does not prefer cereal crops. Such crop rotation can also be effective against soil-borne pathogens. Intercropping green gram and sunflower in cotton was found to reduce leaf hopper population (Rabindra, 1985). Growing sorghum in association with cowpea or lablab reduces the infestation of sorghum stem borer, *Chilo partellus*.

Growing two rows of mustard at 15-day intervals after every 25 rows of cabbage, planted a fortnight later, has protected cabbage against diamondback moth, leafwebber and aphids in Karnataka, India. Similarly 40-day old marigold (which produces predominantly yellow flowers) and 25-day old seedlings of tomato planted in a pattern of 16 tomato rows alternated with one row of marigold in an IPM practice was helpful in protecting tomato against *H. armigera* (Srinivasan and Krishna Moorthy, 1994).

In peanut, it has been observed that leaf miner incidence was less when intercropped with sunflower, lablab, cowpea or pearl millet. Soyabean, when grown as an intercrop with maize, is known to help in fixation of atmospheric nitrogen that can be used by the companion crop, besides reducing the weed population and soil erosion.

Interplanting trap crops to trap insect pests and subsequently destroying the trap crop to save the main crop is another common cultural practice. Castor, the most preferred plant, can be raised as a trap crop in cotton, chilies, and tobacco to reduce damage by *Spodoptera litura*. Similarly, okra can be used as a trap crop in cotton to reduce damage by *Earias* spp.

Certain cultural practices are known to encourage and enhance the activities of natural enemies of insect pests and help in exerting natural control. For example, growing maize or cowpea in cotton enhances the activity of lady bird beetles, chrysopids, and other predators. Pollen of maize helps in retaining the adults of *Chrysoperla* in the main cotton field (Anon., 1992). Intercropping peanut with corriander (4:1) was found to be a good practice to encourage the early larval parasitoid, *Campoletis chloridae* which parasitises *Helicoverpa armigera, Spodoptera litura,* and *Achaea janata.* Intercropping maize with soyabean or green gram enhances the activity of spiders (Sarup, 1987).

C.   NUTRIENT MANAGEMENT

Providing optimum fertilizer and water also helps in reducing pest intensity. Higher rates of fertilizer may enhance crop growth but will also favour harbouring a larger pest population. In rice crops, higher levels of N are reported to increase plant hopper population. Similarly, application of a higher rate of nitrogen and late sowing of cotton resulted in more jassid infestation (Aggarwal et al., 1979). Bollworm incidence was found to be minimal in a cotton crop that received 60 kg nitrogen compared with a crop that received 90 kg/ha. In sorghum, application of a biofertilizer, *Azospirillum*, was found to reduce the shootfly damage by enhancing the production of phenolic substances in the plants (Mohan et al., 1987).

D.   HOST PLANT RESISTANCE

Another important element of IPM is using resistant/tolerant varieties. Host plant resistance helps in suppressing the pest population at low cost, with the least disturbance to the ecosystem and also reduces over-dependence on chemical control. There are many success stories in managing serious pest problems with tolerant and resistant crop varieties. Farmers feel more secure with tolerant/resistant varieties, especially in pest endemic areas (Kenmore, 1991)

Other cultural practices include timing of planting dates and proper spacing. Closer spacing in rice is known to favour brown plant hopper incidence. Solarisation by covering the soil with polyethylene sheet raises the soil temperature, thereby killing plant parasitic nematodes and other noxious soil organisms (DeVay et al., 1991).

III. MECHANICAL CONTROL

The mechanical removal and destruction of pests, their developmental stages and/or the affected plant parts may be a useful tool of IPM in many cases. Removal and destruction of egg masses and first instar larvae of *Spodoptera litura* from the field, collection and destruction of grown-up caterpillars is helpful in protecting the crop. Digging trenches to trap and destroy migrating caterpillars of *Amsacta* spp. in peanut is practiced by many farmers. Such trenches are also dug to trap armyworms in maize/sorghum fields as well as to trap locust hoppers in desert areas. Burning trash and making campfires at the seedling stage of peanut attracts adult moths of *Amsacta* spp. In cotton, clipping off the terminal shoots when the crop is 90 days old reduced oviposition by *H. armigera* considerably (Sundaramurthy and Chitra, 1992). In vegetable nurseries, covering the nursery bed with insect-proof net protects the young seedlings against disease-transmitting vectors.

Installation of several traps in the field is done physically to destroy insect pests. Yellow sticky traps are placed in cotton to attract whiteflies. Yellow sticky traps also attract aphids in sorghum fields. Light traps, pheromone traps and spore traps are used for pest monitoring. Fish meal traps are effective for sorghum shoot fly.

Clipping off the tips of rice foliage containing egg masses of stem borer in the nurseries is a sure method of avoiding the damage in transplanted paddy. Removing the fruits and vegetables affected by pests and destroying them avoids further flare-up of the pest. In many developing countries, hand weeding is still the common method of weed control.

IV. BIOLOGICAL CONTROL/BIODIVERSITY

Biological control is the use of living organisms for regulating the incidence of destructive pests. The predators, parasites, and diseases of pests which we use in biological control are a large component of the world's biodiversity. These natural enemies are of enormous value in IPM for sustainable agriculture, where they can often replace the need for pesticide inputs. They are also of value to the control of invasive alien species which threaten the natural ecosystem (Anon., 1994).

The successful control of cottony cushion scale (*Icerya purchasi*), an introduced pest of citrus in California, by its native predatory beetle (*Rodalia cardinalis*) from Australia in 1888–1889 was the beginning of biocontrol initiatives all over the world. Today many countries mass-breed an egg parasitoid, *Trichogramma* for the control of lepidopterous pests. There are over 100 species of these wasps which differ in their selectivity of hosts. They prevent potential damage to the crop by killing the pest in the egg stage itself, and hence act as a useful component in IPM. Mass-release of *Trichogramma* can be timed based on the occurrence of the pest population which can be monitored through pheromone traps. In China, Russia and India, millions of hectares of maize, cotton, and sugarcane crops are treated with these wasps to augment natural suppression of pest species.

Control of cassava mealybug (*Phenacoccus manihoti*) in Africa by a wasp (*Epidinocarsis lopezi*) is another successful example in recent years. The mealybug is reported to cause up to 80% yield loss, and the annual damage was estimated to be about U.S. $2 billion to this very important food crop of Africa. After several years of searching, the wasp was found in Paraguay parasitising the same species of cassava mealybug. It was mass-multiplied and released in more than 15 African countries, and the pest was brought under natural control. In India, coccinellid beetles, viz., *Cryptolaemus montrouzieri* and *Scymnus coccivora*, are found as effective predators on mealybugs in coffee, citrus and grapes. Coccinellids, chrysopids, and syrphids found to be feeding on soft-bodied insects such as aphids, thrips, and whiteflies in vegetable crops are potential bioagents. A notable example in India is the control of sugarcane leafhopper (*Pyrilla perpusella*) by its indigenous nymphal/adult parasitoid (*Epiricania melanoleuca*). By preventing the crop from being sprayed in 18.63 lakh acres (about 0.8 million ha) and conserving the parasitoid, there was a saving of 40 million rupees, besides avoiding environmental contamination (Banerjee, 1977, Misra and Pawar, 1984).

In apple orchards in India, mass releases of *Trichogramma* spp. against codling moth, predatory beetles (*Chilocorus* spp). and parasitoids (*Aphytis diaspidis* and *Encarsia* sp.) against San Jose scale have been found effective (Sankaran and Manjunath, 1988). Major success has been achieved in controlling coconut black-headed caterpillar by its parasitoids, *Goniozus nephantidis* and *Bracon brevicornis*. Other organisms that are used for biological control are certain nematodes that act as entomo-pathogens, viz., *Romanomermis* spp., which are effective against dipteran pests (flies, gall midges, leafminers) and *Steinernema feltiae* against caterpillars and termites. A protozoan, *Nosema locustae*, is known to affect locusts.

A sucking insect pest from Asia (*Rastrococcus invadens*) became a major pest of mango throughout West Africa in the early 1980s. In 1987, introduction of a specific parasitic wasp from India into Togo reduced the population of this mango mealybug to nondamaging levels, with an estimated benefit of U.S. $3.9 million per year (Vogele et al., 1991).

There are a few examples of controlling weeds through phytophagous bioagents or plant feeding organisms. Control of rubber vine, *Cryptostegia grandiflora*, in Australia through the rust fungus, *Maravalia cryptoslegiae*, and in other parts of the world, including India, control of water hyacinth through *Neochetina* spp., water fern through *Cyrtobagous salviniae*, cactus through a mealybug, *Dactylopius opuntiae*, and lantana through a sap-sucking bug, *Teleonemia scrupulosa*, are other examples. Recent studies indicate that there are a number of natural enemies including insects, mites, nematodes, fungi, bacteria, and viruses that could be used for possible control of more than 28 major weeds in Southeast Asian countries (Waterhouse, 1994).

Biological control of soil-borne plant pathogens in crop plants by way of treating the seeds with fungal bioagents, viz., *Trichoderma viridi*, *T. harzianum*, *T. koningii*, and *Gliocladium virens*, has been successful in crops like pulses, oilseeds, cereals, millets, sugarcane, plantation, and vegetable crops.

In major rice-growing countries in Asia, conservation of bioagents in the rice ecosystem such as spiders, wasps, mirid bugs, waterbugs, damsel-flies, dragon flies, ground beetles, etc. is a vital part of IPM and extension education. The role of birds as predators of many insect pests in croplands should not be undermined.

The above examples show how biodiversity can be used to protect biodiversity. During the last century of research on biological control, an estimated 3000–5000 introductions of natural enemies have been attempted worldwide, involving about 1000 species of bioagents against 200 pest species. One third of these introductions resulted in establishment of these bioagents. Of these, 58% attained partial control and 16% provided complete control of the target pest (Hall et al., 1980). These natural enemies of pests maintain a balance of populations of their prey and ensure coexistence of species by allowing none to become too abundant, which in turn generates the biodiversity in nature. The agro ecosystem around the world has a rich fauna and flora of natural enemies which protect food, fibre, and timber. Their conservation reduces our dependence on agrochemicals and limits the negative effects to environment. IPM involves a reduction in dependence on pesticides and an increased reliance on the action of natural enemies which is free to the farmers. Therefore, conservation and augmentation of beneficial species in the overall ambit of IPM is the preferred strategy to achieve sustainable agriculture.

V.   CHEMICAL CONTROL

Pesticides have played a vital role in developing countries to protect human health, in vector control as well as in sustaining agricultural production to secure adequate food and fibre to their ever-increasing population. However, in view of the well-known hazardous effects to human health, and other nontarget organisms, their use should be carefully considered in order to obtain optimum benefits. In 1988, the global pesticide market was estimated to be over U.S. $20 billion, 16% of the consumption being in the Asian and Pacific regions. Herbicides are widely used in the United States, fungicides in Western Europe and insecticides in the Asian and Pacific regions. While insecticides are used mainly in rice, cotton, and vegetables, fungicides are mainly used in vegetables. The proper and restrained use of safe pesticides should be the major objective of IPM strategy. Pesticides should be used according to need and through observing the economic threshold level (ETL), which is defined as "the population density of a pest which will cause sufficient loss to justify the cost of control." ETL should take into account not only the pest population, but also the population of natural enemies present in the ecosystem. An ideal pesticide is one that will destroy the target pest species with minimum damage to natural enemies and nontarget organisms; it should be inexpensive and degrade fast without leaving residues on the produce. Proper application method and correct dose and time of application play important roles in the success of chemical control. The choice of pesticide is also of great importance. Broad-spectrum pesticides should be avoided. Certain molecules have been obtained from biological species which have high insecticidal property and are considered safe for the environment. Attempts have been made in successfully identifying molecules that affect only a particular stage (damaging stage) of the insect.

Botanical products, particularly neem and its derivatives, are considered as environmentally safe pesticides. More than 1000 plant species having insecticidal properties, 384 species with antifeedant property, 297 with repellent property, 27 with attractant property, and 31 species with growth inhibitory property have been reported (Grainage et al.,1985). The insect growth regulators which are synthetic insect hormones, particularly the benzophenyl ureas which inhibit chitin formation during the moulting process and the juvenile hormone which retains larval characters, interfere with insect metamorphosis and are also host specific. Many plant products which mimic these characters could be exploited as useful tools in IPM.

Use of biopesticides derived from microorganisms, viz., bacteria, viruses, fungi, and protozoans, is another area gaining importance in IPM. The nuclear polyhedrosis viruses (NPV) and granulosis viruses (GV) have a high degree of host specificity, and hence do not affect the nontarget organisms. Bacterial pathogens, *Bacillus thuringiensis* and *Bacillus popillae*, are now commercially available for pest control. *B. thuringiensis* is a widely used biopesticide against insect pests such as *Helicoverpa armigera* and *Plutella xylostella*, which have developed resistance to most of the conventional insecticides. *B. thuringiensis* var. *kurstaki* is widely used against caterpillars and beetle grubs, while *B.t.* var. *israelensis* is most effective against larvae of mosquitoes and blackflies. *B. popillae* has been successfully tested against white grubs. NPV is mostly used against *H. armigera and Spodoptera litura* affecting pulses, oilseeds, vegetables, and cotton in India. Certain entomopathogenic nematodes already described are also considered as biopesticides.

Fungal microbial insecticides are capable of penetrating insect cuticle and multiply inside the body, producing lethal metabolites. Formulations of such entomofungi (*Beauveria bassiana*, *Verticillium lecanii*, and *Metarhizium anisopliae*) have shown promising results against several insect pests. Major success has been achieved very recently in containing coffee berry borer with *B. bassiana* in the Coorg area, India. Other fungal bioagents such as *Trichoderma* spp. and *Gliocladium vireus* are used for seed treatment. Since these microbial pesticides are highly host specific and nontoxic to other organisms, there is a potential to use them in vegetables. Besides, they are safe to the operator.

Locust invasions affect farming communities in about 57 countries from Africa to Asia, and application of toxic chemicals is detrimental to the flora and fauna of the treated areas and causes large-scale environmental contamination. Therefore, biopesticides are being considered for the control of locusts and grasshoppers. Infections caused by the protozoan, *Nosema locustae*; bacteria, *Coccobacillus arcidiorum*; fungi, *Entomophaga (Entomophthora)* spp., *Beauveria* spp., and *Metarhizium* spp., are being exploited besides other insect bioagents. *Bacillus thuringiensis* has not been isolated from Orthoptera, as the low pH of the gut prevents toxic crystals from dissolving. Similarly, NPV is not known to occur in nature in the Acridoidea (Greathead, 1992).

VI.   GENETIC ENGINEERING

Modern biotechnology emphasises the improvement of genetic characters of the cell by exploiting recombinant deoxyribonucleic acid (DNA) to develop improved and modified organisms. It complements plant breeding. By genetic manipulation, there is a possibility of overcoming incompatibilities between species. The incorporation of resistant genes into crop plants renders them less susceptible and vulnerable to injurious organisms, viruses, or the phytotoxic effect of pesticides and more tolerant to stress. Some transgenic or genetically modified plants are already available for cross-breeding into existing varieties of crops. By transferring a natural insecticide gene originating from *Bacillus thuringiensis* into plants, the plants can be made resistant to insect species that feed on them. Such transgenic plants produce a toxin which permeates into the intestine of the feeding insect and kills it. Tomato, potato, and tobacco transgenic plants have been obtained with a built in protection against infectious viruses by incorporating the "coat protein gene" of tobacco mosaic virus (TMV) and potato X virus. By transferring a gene that is resistant to herbicides in the bacterium (*Agrobacterium tumefaciens*) into plants, herbicide resistant plants can be realised (Marshall and Walter, 1994). A novel strategy in development of new biopesticides by developing genetically engineered baculoviruses for pest control is under way (Rajendra, 1994). Insect-specific peptide neurotoxins have been isolated from the venom of South Indian scorpion and incorporated into nuclear polyhedrosis viruses. The recombinant NPV enhances its insecticidal property that is effective against *H. virescence* larvae but nontoxic to blowfly larvae and mice.

VII.   REGIONAL INITIATIVE FOR IPM AND HUMAN RESOURCE DEVELOPMENT

A.   South and Southeast Asia

Rice is the major staple food in Asia. Rice production employs more than 200 million families and the current annual production is estimated to be 478 million tonnes, which is 92% of the world total for the crop (FAO 1990 AGROSTAT). Recent studies by the International Rice Research Institute (IRRI) and the International Food Policy Research Institute (IFPRI) have shown that the expansion of the rice area has almost stopped, and rice yields stagnate or decline under continuous intensive production due to environmental degradation, including factors such as increased pressure from pest outbreaks, depletion or imbalances of soil nutrients and deterioration in water quality. For raising rice production and improving food security in Asia, IPM could play a vital role by empowering the farmers to select, adapt, and apply technologies which are productive, profitable, and sustainable. Research and practice have proven that farmers can conserve biodiversity, increase yields, and make higher profits by using pesticides more judiciously. Restricting the type, timing, and volume of pesticide applications will conserve the "natural enemies" of rice pests, which provide effective biological control in the majority of rice fields. To achieve this, farmers must be able to monitor the plant/pest ecology of their rice crops, and make decisions which are specific to their own fields. Experience has shown that these decisions cannot be packaged as part of extension messages; instead, farmers must learn the underlying principles and make the decisions for themselves (Kenmore, 1991, Kenmore et al., 1984, 1987).

1.   Bangladesh

Rice hispa was considered as the most troublesome pest in Bangladesh. Farmers and extension officers approached the government for aerial spraying to control this pest. However, the government took a major initiative to train the extension person- nel on IPM in rice since 1988. As a result, so far there has been no resort to aerial spraying of pesticides. There have been 2255 extension officers and 309 NGOs trained so far through the FAO Intercountry IPM project. Besides, 91 farmers' field schools were established in 39 districts, and 28,824 farmers received training on IPM. IPM-trained farmers reduced the use of pesticides by 80% compared to farmers without IPM training.

2.   India

India experienced a number of serious pest outbreaks due to misuse/abuse of pesticides. Notable examples are whitefly and bollworm damage in cotton; BPH and leaf-folder damage in rice; and diamondback moth damage in cabbage. Most of these pest outbreaks have been proved to be pesticide-induced outbreaks. Therefore, India adopted integrated pest management as its principle and main plank of plant protec- tion strategy in its overall crop production programme since 1985, with a view to achieving sustainable crop production with optimum cost, preserving the ecosystem in the environment, improving the quality of agricultural produce, and minimising health hazards. A network of IPM centres have been established throughout the country for organising IPM field demonstrations and training farmers and extension functionaries. Over 10,000 demonstrations have been organised so far, and more than 100,000 farmers and 9000 extension officials have been trained in IPM. IPM Dem- onstrations have brought out two important findings, viz. (1) increase in yield in almost all the IPM demonstrations, and (2) reduction in the use of pesticides in IPM fields as compared to non-IPM fields.

The current major initiatives are (1) human resource development through season-long training of trainers and establishment of farmers' field schools (FFS) in every village to make the farmers understand the role of naturally occurring benefi- cial fauna, the built-in compensatory mechanisms in the plant and to analyse the agro-ecosystem, thus empowering them to make their own decisions; (2) removal of subsidies on pesticides; (3) levying excise duty on pesticides; (4) promotion of plant based/neem-based products, etc.

3.   Indonesia

Data available in Indonesia convincingly proved that BPH outbreaks in rice were induced by heavy insecticide use and that resistant varieties only offer a temporary solution if heavy pesticide use is continued. In 1986, the resistance mechanism was broken by BPH and threatened to have a devastating effect on rice production. Realising the mechanisms behind such outbreaks, the Government of Indonesia took the following initiatives:

1. Declared IPM to be the official pest control strategy through Presidential Instruction 3/1986.
2. Banned 57 trade formulations of insecticides for use in rice.
3. Increased the number of field pest observers in rural extension programmes.
4. Trained large number of farmers in IPM.
5. Removed subsidies on pesticides.

Such initiatives have helped in reducing pesticide use by over 50% and increasing rice production by over 12%. Farmers' field training to make them experts on making their own decisions had tremendous impact on the frequency of pesticide application by 0.8 per season by trained farmers, while they achieved higher yields than untrained farmers (Kenmore, 1991).

4.   Philippines

The Philippines have also declared a national policy on IPM since 1986. The previously dominant molluscicides for control of the exotic apple snail (*Pomacea*), triphenyl tin products, have been totally banned since 1991. With mounting evidence of insecticide-induced outbreaks of BPH in Mindanao, adverse health consequences of exposure to the major commercial insecticides, and the problems with molluscicides, the Government prepared the first national policy statement which has been endorsed by the National Economic Development Authority, and an interdepartmental task force established with the Departments of Health, Agriculture, Environment, and Natural Resources and the appropriate university experts (Kenmore, 1991).

5.   Sri Lanka

Despite disturbances in Sri Lanka, IPM is being promoted in rice through training and demonstrations. A recent survey has revealed that IPM-trained farmers had significantly less BPH, significantly more yield, and used significantly less insecticides even when there is no national IPM policy in the country (Kenmore, 1991).

6.   Thailand

A similar situation to the one in Indonesia was seen in Thailand when over 250,000 hectares of rice was affected by serious outbreak of brown plant hopper during 1989–1990. A national conference was held in October, 1990 at which scientists from Indonesia and IRRI presented their experiences and results on BPH and IPM and a policy proposal was made for IPM training which can quickly bring the problem to manageable proportions.

7.   Vietnam

Vietnam also had serious outbreaks of BPH in 1990 and 1991 affecting 400,000 ha of rice. Field study data indicated that pesticides wiped out the egg-feeding predatory bug, *Cyrtorhinus*, water striders that feed on newly hatched BPH and other spiders which kill the older BPH and caused the outbreak. On the basis of these findings and the experiences of Indonesia and Thailand, a policy decision for field training on IPM was initiated in August 1991.

Comparison of rice yields in IPM and farmers' practice in different countries in South and Southeast Asia is given in Table 1.

Increase in the yield and reduced use of pesticides due to IPM in comparison to conventional control practice has also been recorded in many countries on a number of crops (Kenmore, 1993):

Wheat: Holland
Rice: Bangladesh, China, India, Indonesia, Philippines, Sri Lanka, Thailand and
      Vietnam
Vegetables, coconut, oil palm and cacao: Malaysia
Soyabean: Brazil
Sugarcane and cotton: Pakistan
Cotton: U.S. (Texas), Egypt, Sudan and Peru
Coconut: Australia (W. Samoa)

Sustained yields but reduced use of pesticides due to IPM have been recorded on the following crops:

Wheat: Denmark
Vegetables and fruits (apple): Holland
Vegetables: Indonesia, Philippines and Taiwan
Fruits (apple): U.S. (California)

B.   FAO INTERCOUNTRY PROGRAMME FOR IPM IN RICE FOR SOUTH AND SOUTHEAST ASIA

The FAO/Government Cooperative Intercountry Programme for Development and Application of Integrated Pest Control in rice in South and Southeast Asia currently in operation in Bangladesh, China, India, Indonesia, Malaysia, Philippines, Srilanka, Thailand and Vietnam aims to develop and implement IPM training programmes to extension staff and farmers, establish and strengthen IPM programmes and exchange information among countries in support of national policies on IPM.

Over the last 10 years, the FAO Inter-Country Programme (ICP) for IPM has shown that this approach, involving "experiential learning" by farmers, is not only feasible, but it produces additional benefits in other aspects of rice production. The IPM strategy for the region, in simple terms, is to replicate this approach on a large scale.

TABLE 1

Comparison of Yields of IPM and Farmers' Practice (FP)

Country	Treatment	Mean Yield	Period	N	Paired T-test T Value	Paired T-test P	Wilcoyon Z value	Wilcoyon P
All countries	IPM FP	4,904 4,650	1981–91	444	10.8	.0001	–10.0	.0001
Bangladesh	IPM FP	3822 3,340	1989–91	42	7.1	.0001	–4.8	.0001
China	IPM FP	6202 5,593	1989–90	6	9.0	.0003	– 2.2	.0277
India	IPM FP	4,777 4,452	1986–90	18	4.1	.0008	–2.9	.0033
Indonesia	IPM FP	6,031 5,921	1987–90	131	2.4	.0174	–2.1	.0344
Philippines	IPM FP	5,154 5,033	1981–90	91	2.1	.0345	–2.2	.0281
Sri Lanka	IPM FP	3,695 3,044	1986–89	31	7.9	.0001	–4.7	.0001
Vietnam	IPM FP	4,094 3,779	1990–91	117	9.5	.0001	–8.0	.0001

Note: The sample unit is a mean representing at least 15 farmers for each management alternative, season, and location, so that results from over 10,000 farmers' crops are represented.

From Peter Kenmore, FAO Regional Programme Coordinator, Manila. Personal communication.

Farmers could be trained by extension staff to master IPM field skills, apply them in their own fields, reduce pesticide use, increase profits and maintain or increase their yields.

Integrated pest management is now practised by over 500,000 farmers (and known to millions) through the efforts of the FAO Inter-Country IPC Rice Programme and its participating countries. It is anticipated that more than 7,702,500 farmers will be trained in IPM field activity by 1998 (Table 2). IPM is the only large-scale example of farmers empowered to use knowledge of intensive technology and to continue adapting such a technology to changing environmental and economic conditions.

The current status of the impact of IPM in the member countries of the FAO Inter-Country Programme is given in the following tables, which are all based on data from the 1994 F.A.O. G.C.P. Terminal Report.

TABLE 2

Estimated Number of Beneficiary Farmers by 1998

		Fully Trained Farmers	
Country	ICP Funds	Other Funds Supporting National Programmes	Farmers Who Have Participated in at Least One IPM Field Activity
Bangladesh	20,000	200,000	300,000
China	30,000	250,000	250,000
India	10,000	100,000	400,000
Indonesia	20,000	1,500,000	3,000,000
Malaysia	7,500	25,000	25,000
Philippines	70,000	150,000	250,000
Sri Lanka	20,000	100,000	200,000
Thailand	10,000	50,000	200,000
Vietnam	15,000	100,000	400,000
Total	202,500	2,475,000	5,025,000
		Grand Total	7,702,500

From Peter Kenmore, FAO Regional Programme Coordinator. Personal communication.

Country: BANGLADESH

Current Status

Annual rice area	10,600,000 ha
Annual rough rice production	29,400,000 ton
Average rough rice yield	2.77 t/ha
No. of rice farmers	10 million
Annual volume of formulated insecticides	6,500 ton
Current national IPM programme stage	National Policy Constituencies identified and meet Pilot Farmers' Field Schools
No. of IPM trainers	2,000 Government; 200 NGO
No. of IPM trained farmers	3,200

	IPM-Trained Farmers	Untrained Farmers
No. of pesticide applications		
Average pesticide expenditures	147 Taka/ha (4.2 US$/ha)	661 Taka/ha (18.9 US$/ha)
Average yield (1990/1991)	4.3 t/ha	3.64 t/ha
Average yield (all trials, 1989–1991) (n = 42)	3.8 t/ha	3.3 t/ha

Country: INDIA

Current Status

Annual rice area	42,800,000 ha
Annual rough rice production	109,500,000 ton
Average rough rice yield	2.62 t/ha
No. of rice farmers	30 million
Annual volume of formulated insecticides	200,000 ton
Current national IPM programme	Training of first generation of IPM Trainers
No. of IPM trainers	50
No. of IPM trained farmers	50,000

	IPM-Trained Farmers	Untrained Farmers
No. of pesticide applications	0.8 per season	2.4 per season
Average pesticide expenditures	163.5 Rs/ha (9.1 U.S.$/ha)	447.9 Rs/ha (24.9 U.S.$/ha)
Average yield (1991)	5.5 t/ha	5.1 t/ha
Average yield (all trials, 1986–1990) (n = 18)	4.8 t/ha	4.4 t/ha

Country: INDONESIA

Current Status

Annual rice area	10,000,000 ha
Annual rough rice production	43,846,000 ton
Average rough rice yield	4.38 t/ha
No. of rice farmers	15 million
Annual volume of formulated insecticides	9,000 ton
Current national IPM programme	Local government commitments; financial discretionary and regular budgetary
No. of IPM trainers	2,000
No. of IPM-trained farmers	200,000

	IPM-Trained Farmers	Untrained Farmers
No. of pesticide applications	0.8 per season	2.2 per season
Average pesticide expenditures	7349 Rp/ha (3.7 U.S.$/ha)	15307 Rp/ha (7.7 U.S.$/ha)
Average yield (1991)	6.6 t/ha	6.3 t/ha
Average yield (all trials, 1987–1990) (n = 131)	6.0 t/ha	5.9 t/ha

Country: PHILIPPINES

Current Status

Annual rice area	3,525,000 ha
Annual rough rice production	9,600,000 ton
Average rough rice yield	2.72 t/ha
No. of rice farmers	1.6 million
Annual volume of formulated insecticides	9,000 ton
Current national IPM programme	National IPM policy declaration
No. of IPM trainers	11,000
No. of IPM-trained farmers	175,000

	IPM-Trained Farmers	Untrained Farmers
No. of pesticide applications	1/season	2/season
Average pesticide expenditures	363.1 peso/crop (13.4 U.S.$/crop)	705.1 peso/crop (26.1 U.S.$/crop)
Average yield (1990)	4.6 t/ha	4.2 t/ha
Average yield (all trials, 1981–90) (n = 91)	5.1 t/ha	5.0 t/ha

Country: SRI LANKA

Current Status

Annual rice area	735,000 ha
Annual rough rice production	2,200,000 ton
Average rough rice yield	2.99 t/ha
No. of rice farmers	0.9 million
Current national IPM programme	Pilot farmer field schools
No. of IPM trainers	2,800
No. of IPM-trained farmers	87,000

	IPM-Trained Farmers	Untrained Farmers
No. of pesticide applications	0.7/season	2.7/season
Average pesticide expenditures	210.43 Rp/ha (5.3 U.S.$/ha)	693.17 Rp/ha (17.3 U.S.$/ha)
Average yield (1988)	3.3 t/ha	2.7 t/ha
Average yield (all trials, 1986–1989) (n = 31)	3.7 t/ha	3.0 t/ha

Country: THAILAND

Current Status

Annual rice area	10,200,000 ha
Annual rough rice production	20,400,000 ton
Average rough rice yield	2.00 t/ha
No. of rice farmers	2.5 million
Annual volume of formulated insecticides	23,000 ton
Current national IPM programme	Pilot farmer field schools
No. of IPM trainers	25
No. of IPM-trained farmers	500

Country: VIETNAM

Current Status

Annual rice area	5,900,000 ha
Annual rough rice production	19,150,000 ton
Average rough rice yield	3.25 t/ha
No. of rice farmers	10 million
Annual volume of formulated insecticides	11,000 ton
Current national IPM programme	Training of first generation of IPM trainers
No. of IPM trainers	80
No. of IPM-trained farmers	6000

	IPM-Trained Farmers	Untrained Farmers
No. of pesticide applications	0.79 kg ai/ha	1.39 kg ai/ha
Average pesticide expenditures	152,353 dong/ha (17.92 U.S.$/ha)	228,883 dong/ha (26.93 U.S.$/ha)
Average yield (1991)	4.2 t/ha	4.1 t/ha
Average yield (all trials, 1990–1991) (n = 11&)	4.1 t/ha	3.8 t/ha

Country: CHINA

Current Status

Annual rice area	32,900,000 ha
Annual rough rice production	188,300,000 ton
Average rough rice yield	5.72 t/ha
No. of rice farmers	129 million
Annual volume of formulated insecticides	450,000 ton
Current national IPM programme	Training of first generation of IPM Trainers
No. of IPM trainers	150
No. of IPM-trained farmers	47,000

	IPM-Trained Farmers	Untrained Farmers
No. of pesticide applications	2.79 per season	3.50 per season
Average pesticide expenditures	92.25 yuan/ha	138 yuan/ha
	(17.1 U.S.$/ha)	(25.6 U.S.$/ha)
Average yield (1991)	6.4 t/ha	5.9 t/ha
Average yield (all trials, 1989–1991) (n = 6)	6.2 t/ha	5.6 t/ha

Country: MALAYSIA

Current Status

Annual rice area	628,000 ha
Annual rough rice production	1,800,000 ton
Average rough rice yield	2.87 t/ha
No. of rice farmers	0.4 million
Current national IPM programme	Training of first generation of IPM trainers
No. of IPM trainers	25
No. of IPM-trained farmers	2,500

At the regional level, there have been no other programmes carrying out IPM activities similar to those of the FAO Inter-Country Programme. At the country level, IPM has been a component of plant protection projects in a number of countries, including Bangladesh, India, Indonesia, Philippines, Sri Lanka, and Thailand. Under some of these projects, governments established pest surveillance systems with the aim of providing better field information on which to base governments' decisions when to spray. These systems have strengthened the physical facilities and technical expertise of the recipient Ministries to a greater extent (FAO, 1993).

C.   LATIN AMERICA AND THE CARIBBEAN

Despite growing awareness of hazards to human health and the environment, pesticide use in Latin America has been increasing and, by 2000 AD, it is expected that the region will be spending U.S. $3.97 billion on pesticides every year (Belloti et al., 1990). Considering the integration of the region's agriculture with international markets which enforce strict limits on pesticide residues in produce and keeping as models the instances where integrated pest management significantly reduced pesticide use, the International Centre for Tropical Agriculture (CIAT) has promoted wider application of the IPM approach in Latin America.

At CIAT, germplasm with resistance to bean pod weevil, mexican bean weevil, and leafhopper has been developed. Leafhopper resistance could be integrated with cultural practices and action thresholds to control it. However, resistant germplasm is not available for other important pests such as whiteflies, leafminers, and pod borers. Therefore, in the IPM projects in Colombia, Peru, and Ecuador, research focus is given in this direction with the combined action of thresholds, sticky traps, and uniform planting dates, funded by the International Development Research Centre. With active participation from the farmers, adoption of new practices is being measured. Resistance to important fungal and bacterial diseases in beans, such as anthracnose, bacterial blight, angular leafspot, fusarium wilt, and root rots, is also being investigated in addition to cultural practices to manage them. Varieties with tolerance to bean gold mosaic virus have been identified and deployed.

In cassava, a resistance mechanism to thrips, mites, and whiteflies has been incorporated into finished varieties. Bioagents for the control of mites, mealybugs, hornworms, and burrowing bugs have been identified and deployed. Component technology for important diseases of cassava, is available and antagonistic strains of a fluorescent pseudomonad and *Trichoderma* spp. have been commercialized to make them easily available to farmers.

Some efforts are being made in IPM for rice in Latin America and the Caribbean by developing germplasm resistant to blast, hoja blanca virus, and its planthopper vector, *Tagosodes orizicolus*. Emphasis is given for cultural practices, interaction of water and fertilizer management, seeding rate, etc. to limit pathogens. Colombian rice farmers could reduce herbicide application by 30% by using action thresholds. Bacteria have been isolated from seeds of *Arachis pintoi* and *Stylosanthes guianensis* and from phylloplanes of *S. guianensis* which inhibit the growth of fungi causing major diseases of forages, beans, cassava, and rice (Anon., 1994).

D. AFRICA

Reports of the IPM Working Group meeting held at Delhi in May 1994 indicate formation of the NGO action group in Africa and West Africa and and IPM Working Group comprising 10 representatives who would examine the need, potential, and initiative in IPM on vegetables. Madagascar, which recently announced IPM as its national policy, is actively looking forward to implementing the IPM programme. A pilot vegetable initiative is also being considered in Ghana. A 3-year regional pilot project in three West African countries is to be implemented. Senegal proposed pilot projects on vegetables, millets, and rice. The prospects for IPM implementation in Africa are very bright.

VIII. IPM RESEARCH

Many international and national research organisations have reoriented their research trials for evolving simple low cost technology for practising IPM at the farm level: the International Rice Research Institute (Los Banos, Philippines); the International Centre for Tropical Agriculture (Cali, Colombia); the International Wheat and Maize Improvement Centre (El Batan, Mexico); the International Centre for Agricultural Research in Dry Areas (Aleppo, Syria). The International Crops Research

Institute for the Semi-Arid Tropics (Hyderabad, India), the International Institute of Tropical Agriculture (Ibadan, Nigeria), and the West African Rice Development Association (Bouake, Ivory Coast) have already taken up large-scale field testing of their research efforts. These organisations also offer training on IPM to scientists and extension functionaries of the developing countries (Swaminathan, 1991).

IX. IPM NETWORK

The consortium for International Crop Protection (CICP) and the U.S. Department of Agriculture's National Biological Impact Assessment Program (NBIAP) have formed a strategic partnership to assemble and support global information and communication on integrated pest management research, training, technology implementation, and policy development. This international IPM electronic database and communication facility is accessible via INTERNET. Monthly newsletters are published on electronic bulletin boards. Emphasis is given to new developments in IPM and to controversial issues, with opportunities for guest editorials and "letters to the editor." It is a readable forum for the exchange of current ideas and policy debates.

X. SPECIAL CONSIDERATIONS FOR THE PROMOTION OF IPM IN DEVELOPING COUNTRIES

A. IPM AND POVERTY ALLEVIATION

IPM training brings direct financial benefits to the participating farmers, both in the form of savings (by reducing the variable costs of production) and in the form of higher gross returns. Data have shown that trained IPM farmers can reduce the use of pesticide by 50% or increase yields by 10%.

The savings which arise from IPM practices allow farmers to redirect the money, labour, and time which is normally spent on pesticide application to more productive and profitable jobs/inputs. Furthermore, IPM reduces the frequency and severity of pest outbreaks, which otherwise may cause farmers to default on loans, lose their land, and emigrate to urban areas.

In addition to these financial and economic benefits, there are important health advantages associated with the adoption of IPM. The World Health Organisation has estimated that, each year, pesticide poisoning affects 3% of all agricultural workers. This represents 25 million cases of poisoning annually, many thousands of which are fatal. Farmers who participate in IPM training are not only taught how to reduce the use of pesticides, thus removing much of risk of poisoning, but are also taught safety measures which should be adopted if pesticide applications become necessary.

Further benefits come from the educational process itself. IPM training is experiential, meaning that it starts with an examination by the participants of the constraints and opportunities with which they are faced. This is a highly rewarding process, involving group exercises which strengthen the analytical, problem-solving, and communication skills of the trainees. As a result, farmers not only learn how to manage pests more effectively, they frequently also experience a "personal transformation" and enhanced group cohesiveness. This enables them to solve a wide range of other problems and initiate a more fruitful interaction with local government bodies and bureaucracies, such as agricultural extension and credit agencies.

IPM field schools are not designed specifically for low-income farmers. However, the collaboration with NGOs helped to ensure that low-income groups do benefit from some of the training. IPM initiatives are expected to improve income, equity, health, transformative education, and empowerment.

B.   THE ENVIRONMENTAL IMPACT OF IPM

Natural biological control of potential pests is featured in the World Conservation Strategy as an essential ecological process". This process is disrupted or destroyed by the injudicious use of insecticides, which kills the natural predators of potential pests.

The primary impact of IPM is a rapid reduction in insecticide use. This reduction helps to conserve biodiversity, and consequently lessens the incidence of pest outbreaks that would otherwise be caused by the disruption of natural control processes. But the conservation of biodiversity which accompanies the introduction of IPM goes far beyond the predators of rice pests; over 20 years of international research in Asian rice and the experiences of hundreds of thousands of rice farmers have shown that insecticides kill and injure people, fish, snails, shrimp, birds, and domestic animals. Furthermore, insecticides often flow from rice fields into fresh and sea waters, threatening inland and coastal aquaculture.

Farmers trained in IPM learn to manage an ecosystem, not only a farm enterprise. This means managing three trophic levels, not only one (rice) or two (rice and "pests"). Conservation of biodiversity is not a side-effect of IPM. It is an enabling objective adopted by the participating farmers, which they are taught to pursue in a conscious and systematic manner, with over 2 million person days of field training in Indonesia alone. This makes the rice IPM programme the largest field ecology training programme ever attempted. The results in Indonesia, and in other member countries on a smaller scale, have shown that farmers can drop their insecticide frequencies from 2.2 applications per season to 0.8 (i.e., most farmers did not apply any insecticides), while maintaining or increasing yields. This represents a massive reduction in the amount of toxic chemicals entering into the rural environment.

C.   IPM TRAINING FOR FARM WOMEN

Women actively participate in the agricultural sector. Farm women are equally exposed to pesticide toxicity as they do share with men in handling, storage, and preparation of pesticides. They are equally exposed, like men, to air, soil, and water contamination and consumption of food contaminated with pesticides. During storage of food grains and seeds, pesticides are mostly handled by women at the domestic level. Women are more prone to exposure to pesticides in operations like sowing, weeding, and harvesting. In India, women are employed in large number (81.23% of 44.97 million total women in the workforce compared to 65.50% of a 177.55-million workforce in the agricultural sector) in mills and factories where they are exposed to pesticide residues. Since literacy in rural women is low, they are more prone to contamination out of sheer ignorance (Chopra, 1993). Therefore, educating these farm women is another special area of consideration in IPM strategy.

XI. SUMMARY

With the global concern on the adverse effects of pesticides on human health and environment, there is a need to formulate an alternative strategy in pest management. Integration of cultural, mechanical, biological, and chemical control is the best method of managing pest problems, rather than continuing our reliance on synthetic chemicals alone. Genetic engineering, biopesticides, plant products, and molecules that are safer for nontarget organisms and the ecosystem would supplement and complement IPM strategies. Human resource development in the developing countries will play a vital role in reducing pesticide usage and conserving biodiversity.

IPM is bound to bring happiness among farmers.

How can we do this ?
How long will it take ?
How much will it cost ?

ACKNOWLEDGMENT

The information presented in this Chapter is largely based on published articles and reports and personal communications from a number of IPM experts. We are particularly grateful to Dr. Peter E. Kenmore, FAO Regional Programme Coordinator in South and Southeast Asia, Manila, Philippines, who is spearheading the IPM movement in South and Southeast Asia, for his continuous guidance and support in preparing this Chapter. We are grateful to Dr. S. Ramaswamy, FAO, IPM Specialist, Bangladesh for his support and encouragement. It is a pleasure to acknowledge the support rendered by many colleagues, especially Mr. K. C. Vermani, Mr. J. Mitra, and Mr. S. Krishnamoorthy in converting the manuscript into presentable form.

REFERENCES

Aggarwal, R. A., Wankhede, M. P., and Katiyar, K. N. 1979. Impact of agronomic practices in the population of jassid *(Amrasca devastans* Distt.) in cotton. *Cotton Fibres Trop.* 34:375–378.

Anonymous. 1992. Proceedings of National Workshop on IPM for harmonisation of package of practices. 29–30 June, 1992, CPPTI, Hyderabad (India).

Anonymous. 1994. A Regional Initiative for Integrated Pest Management in Latin America and the Caribbean: How CIAT can contribute. Paper prepared by Pest and Disease Management Scientific Resource Group. International Centre for Tropical Agriculture, Columbia, 17 pp.

Banerjee, S. N. 1977. Plant protection: past, present and future. *Indian J. Plant Prot.* 5(1):1-2.

Bellotti, A. C., Cardona, C. and Lapointe, S. L. 1990. Trends in pesticide use in Colombia and Brazil. *J. Agric. Entomol.* 7(3): 191–201.

CAB International. 1994. Using Biodiversity to protect Biodiversity — Biological Control, Conservation and the Biodiversity Conservation, Oxon, U.K., 14 pp.

Chopra, V. L. 1993. Keynote address. Proceedings of the National Workshop on Women and Pesticides. New Delhi, November 24–26, 1993, 4–8.

DeVay, J. E., Stapleton, J. J., and Elmore, C. L. 1991. Soil Solarization, Food and Agriculture Organization of the United Nations, Rome.

Food and Agriculture Organization, Agro Stat 1990.

Food and Agriculture Organization, Project Document, F.A.O./Government Cooperative Programme, April, 1993.

Food and Agriculture Organization, G.C.P. Terminal Report, 1994.

Grainage, M., Ahmed, S., Mitchel, W. C., and Hylin, J. H. 1985. Plant species reportedly possessing pest control properties — An EWC/UH Database Resource Systems Institute, East West Centre, Honolulu, Hawaii.

Greathead, D. J. 1992. Natural enemies of tropical locusts and grasshoppers: their impact and potential as biological control agents. In *Biological control of locusts and grasshoppers.* Ed.: C. J. Lomer and C. Pricon. Proceedings of a workshop at International Institute of Tropical Agriculture, Cotonou, Republic of Benin, April/May, 1991. CAB International and International Institute of Tropical Agriculture, Cotonou, 105–154.

Hall, R. W., Ehler L. E., and Bisabri-Ershadi, B. 1980. Rate of success in classical biological control of arthropods. *Bull. Entomol. Soc. Am.*, 26:11–114.

Jayaraj, S. 1992. Integrated Pest Management. In *Pest Management and Pesticides: Indian Scenario.* Ed. B. V. David. Namrutha Publications, Madras, 7–16.

Kenmore, P. E., Carino, F.O., Perez, C.A., Dyck, V. A., and Gutterrez, A. P. 1984. Population regulation of the rice brown plant hopper (*Nilaparvata lugens* Stal.) within rice fields in the Philippines. *J. Plant Prot. Tropics* 1(1): 19–37.

Kenmore, P. E., Litsinger, J. A., Bandong, J. P., Santiago, A. C., and Salac, M. M. 1987. Philippines rice farmers and insecticides: thirty years of growing dependency and new options for change. In Tait, J. and Napompeth, B. (Eds.). *Management of Pests and Pesticides in Developing Countries.*

Kenmore, P. E. 1991. Indonesia's Integrated Pest Management: A model for Asia. How rice farmers clean up the environment, conserve biodiversity, raise more food, make higher profits. Manila: Food and Agriculture Organization. 56 pp.

Kenmore, P. E. 1993. Personal communication.

Marshall, G. and Walker, D. 1994. *Molecular Biology in Crop Protection.* Chapman and Hall, London. pp. 283.

Misra, M. P. and Pawar, A. D. 1984. Use of *Epipyrops melanoleuca* Fletcher (Lepidoptera, Epipyropidae) for the control of sugarcane pyrilla, *Pyrilla perpusilla* (Walker) (Hemiptera, Fulgoridae). *Indian J. Agric. Sci.* 54(9) 742–750.

Mohan, S., Jayaraj, S., Purushothaman, D., and Rangarajan, A. V. 1987. Can the use of *Azospirillum* biofertilizer control sorghum shootfly? *Curr.Sci.* 14:723–725.

Oudejans, J. H. 1991. Agro-pesticides — Properties and Functions in Integrated Crop Protection, United Nations, Economic and Social Commission for Asia and the Pacific, 329 pp.

Rabindra, R. J. 1985. Transfer of plant protection technology in dry crops. In *Integrated Pest and Disease Management*. Ed. S. Jeyaraj. Tamil Nadu Agriculture Univeresity, Coimbatore, 377–383.

Rajendra, W. 1994. Genetically Engineered Baculoviruses for Pest Control (personal communication).

Sarup, P. 1987. Insect pest management in maize. In *Plant Protection in Field Crops*. eds. M. Veerabhadra Rao and S. Sithanantham (Eds.). 105–112.

Sankaran, T. and Manjunath, T. M. 1988. Biological control. In *Forty years of Plant Protection in India 1946–1986*. D. Bap Reddy (Ed.). Plant Protection Association of India, Hyderabad, 75–83.

Srinivasan, K. and Krishna Moorthy, P. N. 1994. IPM on cabbage, cauliflower and tomato using trap crops. Paper presented at the National Workshop on Non-Pesticidal Approach to Pest Management — A New Direction 20-22nd Sept., 1994. National Academy for Agricultural Research and Management, Hyderabad, 1–7.

Sundaramurthy, V. T. and Chitra, K. 1992. Integrated pest management in cotton. *Indian J. Plant Prot.* 20:1–17.

Swaminathan M. S. 1991. From Stockholm to Rio De Janeiro, The Road to Sustainable Agriculture, Monograph No. 4. M. S. Swaminathan Research Foundation, Madras, 5–6.

Vogele, J.M., Agounke, D., and Moore, D. 1991. Biological control of the fruit tree mealybug, *Rastrococcus invadens* Williams in Togo: a preliminery sociological and economic evaluation. *Trop. Pest Manage.* 37:382–397.

Waterhouse, D. F. 1994. Biological Control of Weeds: Southeast Asian prospects. Australian Centre for International Agricultural Research, Monograph No. 26, 302.

12 Future Considerations

Muthukumaran Gunasekaran, Suresh Gunasekaran, and Darrell Jack Weber

We live in a world in which millions are dying of hunger and almost half of all food production is being destroyed by a plethora of pests and diseases. A wide array of organisms including insects, bacteria, fungi, nematodes, viruses and weeds are responsible for this destruction. Crop plants have many natural defense mechanisms against these pests, like morphological barriers such as thick cuticle and waxy layers, and physiological barriers such as phytoalexins and toxins, but these mechanisms are being overwhelmed by the onslaught of many different pathogens. In order to protect these crop plants, an equally diverse array of pest and weed management strategies have been developed and have had varying levels of success.

The majority of these strategies have involved the use of chemical pesticides and herbicides and they have been widely successful. This can be attributed to the fact that a large number of chemicals have been developed that are effective against a wide range of pests and diseases. Moreover, these chemicals are very economical, both to mass produce and test, as well as for the consumer to purchase. Finally, the chemicals are easy to apply and can be stored over long periods of time.

However, in recent years, we have begun to see some rather serious repercussions arising from the extensive use of chemical pesticides and herbicides. It is now apparent that after repeated chemical use, many pests have developed a resistance to the chemical agents. These resistant pests are much more difficult to control than the original organisms, and in many cases cannot be controlled by chemicals alone. It is often overlooked that chemicals have in many cases killed natural predators of pests, destroying natural mechanisms of controlling pest populations.[1] Moreover, many chemical pesticides have been shown to pollute water sources, destroying basic levels of the food chain, as well as soil microflora. In some cases, the pesticides have contaminated the very crops that they were intended to protect. In addition to pollution, many chemicals have been linked to serious health problems in humans. These many disadvantages have led to considerable public mistrust of chemical pesticides.[2] This has prompted both the government and the private sector to look for safer ways of using chemical pesticides and, more importantly, to alternatives for solving the problems.

One promising strategy in controlling pests and diseases of plants is the use of biological methods. It is appealing because it can be used both as an alternative to chemical pesticides and herbicides and in conjunction with them where necessary. Biological control of plant pathogens typically consists of antagonistic strategies (antibiosis, competition or exploitation by direct predation or hyperparasitism) and induced resistance to plant pathogens in the host. Generally, classical strategies of biological control have met with limited success. In many cases, they have been effective, but they have been slow, inconvenient, or uneconomical. However, recent advances in molecular biology have the potential to improve current biocontrol methods. It has been long known that nature has its own ways of controlling plant pathogens, both by resistant plants and natural enemies. By using biotechnology, these natural control mechanisms can be manipulated and improved to better protect crops against pathogens. Work has already begun on strategies to protect a wide variety of crop plants, from tomato and corn to cotton and flax.[3] Much work has already been accomplished, but a great deal more remains to be done or for that matter even begun.

Insects, the most abundant organisms on the earth, are the primary pests of crop plants. They are doubly destructive on plants not only because of the damage they inflict, but also because they are vectors for other microbial pathogens. Consequently, crop plants have many natural mechanisms of defense against insects. Several of these defense mechanisms are physiological barriers. Many crop plants produce phytoalexins, or natural plant antibiotics, that behaviorally deter insects from feeding upon them,[4] or are toxic to insects, slowing their maturation or killing them.[5] Other natural defense mechanisms of plants include morphological barriers that make it difficult for the insect to penetrate the crop plant.

Classical biological control strategies have tried to maximize these beneficial natural defense mechanisms through cross protection. Although the protection was not always complete and the mechanism of protection was not fully understood, cross protection was the primary biocontrol strategy. In many cases it has been successful; however, cross protection does have some serious deficiencies. Most significantly, the process is very slow and tedious. Additionally, the pathogen may adapt in the interim. Moreover, the protection is not always steadily inheritable and can only be transferred between plants of similar species. Another classical approach is crop rotation, which is time consuming and not particularly effective without other accompanying strategies.

Like the classical approach to producing insect-resistant crop plants, the molecular biological approach tries to improve upon the natural control mechanisms of crop plants, but the similarities end there. The molecular approach seeks directly to produce resistant plants by gene manipulation. Many strategies insert new genes for phytoalexin production or manipulate the existing genes so that phytoalexins are produced at times which are more beneficial for the plant. The methodology for these approaches has been reviewed extensively.[6] Central to this type of work is *Agrobacterium tumefaciens*, a bacterium with the remarkable ability to move a portion of its own DNA into a plant cell during infection. Studies have been reported using this bacterium to insert foreign genes into plants.[7,8] The resulting plants incorporated the inserted genes into their own genomes and thus acquired new, stable and inherited traits. There are many promising genes that could be inserted into plant genomes to confer resistance. One thoroughly studied gene is responsible for the production of the delta endotoxin in *Bacillus thuringiensis* (Bt endotoxin), which has proven to be effective against the larvae of lepidopteran pests, perhaps the most destructive insects in the world. The Bt endotoxin, which has been safely used as a pesticide, has been discovered to be the product of a single gene, which has been isolated and characterized.[9,10] This gene has been successfully inserted into tobacco[11] and work is underway in other beneficial plants.[12] Another approach is the modification of existing genes. When damaged by an insect, plants naturally produce oxidative enzymes, which render unutilizable many nutrients needed by the insect and at the same time hinder its ability to detoxify natural and synthetic toxins. One possible suggested strategy is manipulation of the gene responsible for production of oxidative enzymes, so that the enzymes would be produced before insect attack, as a preventive mechanism.[13]

There are advantages to engineering resistant plants over chemical pesticides or classical biocontrol strategies. The plants would always be resistant regardless of changes in the weather as well as at times when a pesticide could not be sprayed. Additionally, plant parts that are difficult to reach with sprays, such as roots, shaded lower leaves, and new growth would be protected. In crops that presently require heavy insecticide use like cotton, the cost of using resistant plants will be comparable to the current cost of repeated chemical insecticide application. Furthermore, it typically costs much less to develop an insect-resistant crop line than an average chemical insecticide.[12] An overwhelming advantage to host resistance schemes is the low risk of environmental contamination or human health problems, because there is no risk of genetic variation or release of toxic compounds. There should be no toxicity to nontarget organisms because the only exposed organisms are those attempting to feed on the plant. There is always some risk that the genetically engineered plant may respond differently in the environment, but it is much simpler to test the plants than chemical pesticides and is less time-consuming than traditional cross-breeding methods.[12]

There are also possible drawbacks to current molecular approaches in engineering resistant plants. One very valid reservation is that insect resistance may be developed to phytoalexins. Specifically, the molecular mode of action of the Bt endotoxin on many insects is not fully understood; it is still unknown how the bacterial components interact with the insect tissue and exactly what prompts the different toxicological responses in different insects. It is fully possible that some insects could develop a resistance to this or other toxins, much as resistance is developed to chemical pesticides.[14] This point only highlights the need for more study on the mechanisms of action of different toxins, the different pathways of metabolite production in plants, and study of the plant genome in general.[15]

Another approach in controlling insects involves employing other organisms that are natural enemies or insect predators. Classical strategies have increased the number of natural pest predators or transplanted a predator from another environment to control pest populations. These strategies have had only limited success because the predators are frequently unable to survive and reproduce in the new environment or they lose their virulence. Molecular approaches focus on tailoring predators at the gene level through recombinant DNA technology, so that the predator will survive and reproduce under new environmental conditions and remain sufficiently virulent to control target pest populations.

Molecular approaches, like classical ones, utilize a wide selection of insect predators. One of the most intriguing approaches is the use of insects that both attack and compete with predators. These approaches manipulate the predator at the gene level to increase its rate of reproduction, ensuring that there are a sufficient number of hyperparasites or predators to control the pests. In addition, the genes are altered so that the progeny can survive in harsh conditions. The life cycle of the progeny is adjusted in a manner that most effectively controls pest populations. All of these individual techniques begin with the isolation of the appropriate gene which is responsible for the desired trait. Later, this gene can be transferred into the selected predatory organism. Other approaches involve releasing insects that compete for basic resources needed by the pest. By genetic engineering, these insects can be given a competitive advantage over the pests. Finally, as done in *Drosophila melanogaster*, genes may be inserted into the germline to produce sterile males, which compete for existing females, thereby reducing the population.[16]

Viruses are also a potentially useful group of insect predators. Traditional use of baculoviruses as viral insecticides has proven to be an effective, but slow process.[17] Recent studies have indicated that by introducing appropriate foreign genes into the baculovirus genomes, pathogenicity and insecticidal effectiveness may be increased.[18] These include genes for production of insect-specific toxins, hormone production, metabolic enzymes and growth regulators.[19,20] The nuclear polyhiedrosis viruses (NPVs) have also received considerable attention as pest control agents.[21] These baculoviruses are extremely safe, are known to infect only invertebrates, and are not phytotoxic.[22-24] Moreover, the viruses have great commercial potential as they can be formulated to optimize storage, wetting, suspension, flow and spraying characteristics.[23] The virus becomes active upon ingestion and reproduces during the development of the disease.[25] Another strategy is to make a recombinant baculovirus of increased toxicity by introducing appropriate foreign genes into the highly expressed polyhedron gene site.[23] Polypeptide toxins that block neuronal function and the important peptide hormones are also currently being studied. Most of these toxins affect only insects. Infection with such recombinant viruses may cause direct toxicity, alter behavior or arrest insect development. The onset of these effects may be much faster than the pathology caused by the original parent virus. If the appropriate gene for insect toxin and the promoter can be isolated and introduced into a baculovirus, the resulting recombinant virus will be a rather effective biocide.

One advantage in the use of baculoviruses is that there is low risk for environmental contamination because they are extremely host specific. Although it is possible to engineer a more generic virus, one must be careful not to make it too generic, because the more generic it is, the greater the potential threat to animals and humans. There are additional concerns with the use of baculoviruses. Releasing a genetically enhanced virus may displace naturally occurring viruses, which could have a serious impact on the environment. There is always the risk that a virus could have unforeseen deleterious effects once in the environment. Should this happen, it would be extremely difficult to remove the virus from the environment, because it would have such high survival capabilities. A potential solution to this problem could be the engineering of strategies that limit the survival of baculoviruses. Another concern is that the release of a modified organism might cause genetic transfer to nonmodified organisms or genetic reassortment. It is extremely important that engineered viruses be genetically stable.[17] Most currently used baculovirus insecticides are slow in action, from several days to a few weeks, depending on the virus, the host and environmental conditions. This is indeed a great disadvantage compared to chemical insecticides and other natural predators of insects, which act in a matter of hours. Further work is called for in inserting genes into the viral genome, such as insect hormone genes, which will make it possible for the baculovirus insecticide to act more quickly and efficiently.[26]

Another promising frontier in the control of insects is the use of bacterial insecticides. As noted in the previous discussion, one commonly studied and utilized bacterial insecticide is *Bacillus thuringiensis*. It produces large insecticidal crystal-protein inclusion bodies (toxins) during sporulation, which are lethal to many lepidopteran species and to selected species of dipterans and coleopterans; a natural advantage to these proteins is that they are harmless to animals and plants.[1] *B. thuringiensis* is not the only bacterium with insecticidal properties. Over one hundred entomopathogenic bacteria have been identified and reviewed.[21] These belong to the families Bacillaceae, Micrococcaceae, Lactobacillaceae and Pseudomonadaceae. Many of these bacteria have also been shown to be effective in the control of vector insects because of their virulence at low levels and a relatively broad host range which does not include nontarget organisms. Other advantages include their low production costs and the ability to easily mass culture them.[27] Further work needs to be done to modify the bacteria so that they are more host specific, producing toxins that are effective against a particular insect, and to find more virulent strains of such bacteria. Two molecular techniques that should be investigated further and utilized in making more virulent strains are selective plasmid curing and conjugal transfer. The conjugal transfer approach involves the mating of strains with different insecticidal genes and arriving at a strain that contains both genes. The plasmid curing approach seeks to isolate bacterial variants that have lost one or more plasmids carrying genes of low toxicity and to maximize the expression of the remaining high activity toxin genes. With these genetic improvements, bacterial insecticides show great potential for commercial application.[28]

Other strategies with considerable potential involve fungal insecticides. Fungi have long been of interest for the biological control of pests, because of their unique mode of infection and their ability to create epizootics.[29] Most currently marketed mycopesticides have shown little success because of their reputation for unreliability. Many potential mycopesticides have low virulence or efficacy and all are slow acting compared to chemical pesticides.[30,31] A substantial number are unable to survive and control insects in the field; fungi often require high humidity to survive, and can be sensitive to UV light, temperature, and currently used pesticides (especially fungicides).[32–34] Even when fungi are identified which are attractive as potential mycopesticides, production and formulation has been problematic. Some have yet to be grown *in vitro* and others have been shown to lose virulence during propagation and have limited shelf lives.[35,36]

Many new developments in molecular biology offer exciting new ways to remedy problems with myco-insecticides. Recent studies have proposed genetically engineering fungi to increase virulence and efficacy for a specific host.[37,38] DNA transformation systems have been recently developed that would allow specific genes corresponding to desirable traits to be isolated and modified to produce enhanced organisms.[39] Many potential biocontrol fungi have also been shown to produce phytoalexins.[40] The potential of the molecular approach to phytoalexin production must be preceded by study of the molecular interaction between fungi and target insects.

One remarkable advantage of fungal insecticides is their mechanism of action. Recent studies indicate that hundreds of species of entomopathogenic fungi have been identified that do not have to be ingested to kill insects.[23,41] Fungal spores that adhere to the insect surface germinate and send out hyphae that penetrate the cuticle. These hyphae then invade the hemocoel and cause death either rapidly, possibly through the production of complex toxic metabolites, or more slowly because of vast hyphal proliferation and physical disruption of organs. However, myco-insecticides still have poor storage capabilities and are very sensitive to the weather and climate in which they are used. Further work must be done to engineer fungi that can produce beneficial toxins and survive in the appropriate environment.

Another group of insecticides that show a great deal of promise, but have not received much attention in the molecular approach, are nematodal insecticides. There have been numerous reviews of entomopathogenic nematodes and their application in biological control.[42–45] Nematodes, although morphologically very simple, have exploited a wide range of diverse habitats. They have the potential for controlling various organisms: spiders, leeches, annelids, crustaceans, mollusks and many insects. Many nematodes work together with bacteria to control pest populations. The nematode attacks and penetrates the host in the intersegmental areas and the bacterial cells voided from the nematode's intestine into the hemolymph propagate and kill the host by septicemia within 48 hours. The nematodes feed on the bacterial cells and host tissues, produce two or three generations and emerge to search for new hosts. This rapid mortality rate permits the nematodes to exploit a range of hosts that spans nearly all insect orders,[46] a spectrum well beyond the range of any other microbial control agent. For example, studies have shown that steinernematid and heterohabditid nematodes attack a wide spectrum of insects in the laboratory where host contact is assured, environmental conditions are favorable, and no ecological or behavioral barriers to infection exist.[47] Moreover, one nematode species, *Steinernema carpocapsae*, can effectively control field populations of fungus gnats, artichoke plume moths, cutworms, sod webworms, and strawberry and citrus root weevils,[48] a truly wide spectrum of insect hosts.

Success in using nematodes will not only depend on making the nematode more virulent, but ensuring that the organism can survive in a harsh environment. Further study is needed on the behavior of the organism during infective stages. The use of nematodes for biological control shows great promise since they can be grown easily. However, the reasons for success or lack of success in controlling insect pests, particularly in the soil environment, often remain unknown, underscoring the need to obtain basic information on the biology, behavior, ecology and genetics of these nematodes. Understanding behavioral patterns and genetics will enhance production of the most adaptive species for insect control in the field.

Although insecticidal nematodes are beneficial, plant parasitic nematodes cause major damage in agricultural systems, causing losses worth billions of dollars.[49] Classical strategies emphasized crop rotation and cultivar resistance, but these have had only isolated success.[49] Chemical nematicides have not been particularly effective, because nematodes seem to develop a resistance rather rapidly. This has caused many to use multiple, different, chemical nematicides with little overall effectiveness and at great cost to the environment. There are many natural predators to nematodes such as the different varieties of nematophagous fungi, and they work in many ways. Some capture nematodes with simple, undifferentiated hyphae; others use very specialized trapping devices.[50] Additionally, fungi may also produce toxins that paralyze or kill nematodes prior to penetrating their cuticle.[51] Recent studies have found that toxins from fungi are extremely harmful to nematode eggs and young nematodes, killing them in some cases,[52] and causing premature hatching and malformation in others.[53] There have been many studies which have found that toxins from fungi adversely affect the reproductive capacities of female nematodes.[54] The interaction between these fungi and nematodes is very complex, being affected by soil pH, moisture, temperature and available nutrients.[55] Fungi can be applied quite effectively, but they are slow growing and require very specific nutrients in large amounts.[56] Nematodes have also been found to be susceptible to bacterial infections. The bacterium, *Bacillus penetrans*, is an effective biocide and has a life cycle which suits predation on nematodes. The germ tubes penetrate the cuticle and proliferate in the body cavity, killing female nematodes. Bacteria are also promising producers of nematicidal toxins.[57] Because of their nematicidal action and their ease of manipulation for production of toxic compounds, bacteria should receive more attention.[56] Studies[58,59] have demonstrated that nematodes are susceptible to viruses. Nematodes, including those parasitic to plants, have also been noted to prey on their own populations,[60] but very little is known about many of these subjects.

Although much work has been done on the biological control of nematodes, very few of the recent advances in molecular biology have been applied. As proposed by Stirling et al.,[49] one possible molecular strategy is to manipulate the genes of nematicidal fungi or bacteria to overproduce chitinase or collagenase to break down nematode eggs. Other strategies engineer fungi and bacteria to produce nematicidal toxins, as well as deriving more lethal toxins using plasmid curing and conjugal transfer techniques. Another potential strategy is the engineering of nematicidal agents with rhizosphere competence and other mechanisms that allow them to survive in the environments necessary for nematode control. These proposals are highly speculative and much work must be done to determine how molecular tools can be applied to the biocontrol of nematodes.

Fungal diseases are known in virtually all natural plant populations and are a major cause of reduction in agricultural and horticultural yields. In most plant protection systems, breeding for fungal pathogen resistance is a primary concern.[61] One fungal control strategy involves the use of mycoparasitism, where predatory fungi are used to control plant parasitic fungi.[62,63] An example is mycoparasitism of *Rhizoctonia solani* by *Trichoderma harzianum* or *T. viride*. *R. solani* attacks a wide range of plants, causing seed decay, damping off, root and crown rot, stem canker etc., and its control is of major importance.[64] The advantage to this method is that, frequently, the predatory fungi are able to survive and reproduce in the same environment as the phytopathogenic fungi; however, in many cases the predatory fungi may also injure the crop plant. The application of molecular techniques to this field has been sparse. Additional genetic engineering studies of fungi are important to make them very target specific and avoid crop plant damage. Another promising area for investigation is hypovirulence. In some highly virulent pathogens, there have been discovered less virulent strains. If through recombinant DNA technology the gene for the hypovirulence could be isolated and replicated in the general population, it would be most beneficial. Research has already begun in this area and shows great promise. For example, chestnut blight is caused by *Endothia parasitica*. A recent review[65] discussed the existence of strains of this organism which are hypovirulent to normal strains. To make this technique truly successful, the hypovirulent strain must be given a competitive advantage over more virulent strains. The creation of any hypovirulent strain through genetic engineering must be followed by extensive testing before use in the environment to ensure that the engineered organism is indeed genetically stable.

Viruses are one of the major etiologic agents of plant diseases. They are effective pathogens because of their diversity, their ability to reproduce quickly, and their ability to avoid the natural, morphological plant barriers through their small size. Classically, controlling viruses has been accomplished by breeding plants that are resistant or that produce phytoalexins which prevent viral infection. Because of the small size of viruses, there have been very few antagonistic strategies developed. The molecular approach has been widely applied in the area of plant protection from viruses. Many plants have recently been developed that express genes from other plants, bacteria and viruses. One approach that has received much attention is the expression of virus sense or antisense genes in plants to confer viral resistance. The basis of this approach is the classical principle of cross protection, whereby plants infected with a mild strain of a virus do not develop severe symptoms when challenged with a severe strain of the same virus.[66] Furthermore, the molecular approach now enables us to transform and regenerate plants that express genes of many plant viruses. Studies indicate that coat genes have the broadest applicability because they confer specific resistance to the corresponding virus. Studies of other coding and noncoding viral sequences suggest that additional approaches to resistance can be developed.[67] Recent work has demonstrated that the expression of sequences encoding the coat proteins of alfalfa mosaic virus, potato virus X, and cucumber mosaic virus (CMV) provides disease resistance in a number of transgenic plants. Several mechanisms were proposed to explain how the antisense RNA transcript might function in protection of the transgenic plant;[68–70] nonetheless, further study is warranted in this area because this technique shows great potential for success in controlling a large number of plant pathogens.

Much attention has been given to the control of pests and the threat they pose to plants; however, weeds are a significant if not equal threat to crops. They compete for the nutrients of crop plants, harbor pests and pathogens, and produce harmful toxins. They are also particularly difficult to eradicate without harming the crop plant in the process. Some effective chemical herbicides have been developed against particular weeds; however, as previously stated in this chapter, these herbicides have not only been harmful to the weeds, but they have also polluted the soil. The control of weeds by biological methods has been widely reviewed.[71–74] Although fungi, bacteria, nematodes, mycoplasmas and viruses cause plant diseases, there has been only limited use of pathogens other than fungi.[78] The classical biological control strategy for weeds relied upon inoculating fungal agents (mycoherbicides), which would then perpetuate themselves in the environment, controlling the weeds. Mycoherbicides are highly virulent and specific against particular weeds. They are produced for inoculum purposes in artificial culture, and later applied like chemical pesticides to the targeted plants. The disadvantages of current mycoherbicides include absence of sporulation and formulation technology, loss of virulence in culture, and excessively narrow host potential. Furthermore, there has been no effective method for production and storage of these biocides. Fungi are unable to adequately survive and reproduce in the environment where applied.

Although many barriers exist to the development of effective bioherbicides, molecular techniques have been rarely used to address these issues.[75] The molecular approach uses recombinant DNA technology to produce strains of pathogens that are more virulent, produce a wider array of phytotoxins, and can survive and reproduce under harsh conditions.[76,77] For example, a recent study has shown that genetic engineering techniques can create fungi that are resistant to chemical pesticides and can control weeds in the field.[78] Another study has indicated that plants can be engineered to produce toxins that will disrupt the metabolism of specific weeds.[79] Many of these bioherbicides can be engineered to be more host generic or more usable while not interfering with other agents being used at the same time. However, making them more generic does pose more of a risk to the environment and consequently should be tested.[80] The first step in the successful use of molecular techniques is to understand the interaction between host and pathogen. Initial studies in this area are promising, but further research is necessary to determine the true effectiveness of molecular biological strategies in engineering herbicidal agents.

Fundamental to the application of molecular techniques in the biological control of plant pests and diseases is understanding, on the molecular level, the complex relationships between host, pest and predator. In the past two decades, many biotechnological methods have been rapidly developed; technology and procedures that were unthinkable 10 or 20 years ago are now being more frequently applied. To develop effective biological controls, identification of etiological agents and their modes of action, quick and accurate plant disease diagnosis and new production methods are crucial. As described in the introduction, development of new techniques such as PCR, ELISA, RIA and RAPD have been remarkable accomplishments, making identification and diagnosis a rapid and accurate process. Gene identification, cloning and expression to produce antibodies and phytoalexins now offer an infinite range of biocontrols that were not feasible only a short time ago. These examples only serve to highlight the central part that research and development of new molecular biological techniques play in the development of effective biological controls. It is important not only to use currently available biotechnological tools in new applications, but also to develop new and more effective tools for the future.

If biological control is to receive attention in the marketplace, then the techniques that are developed must posess some basic commercial criteria. It is important that the biological control agent be economical to produce. The primary stumbling block in the current development and use of many control methods is the high cost of testing followed by production. With increasing regulations and higher testing standards being placed on chemical pesticide companies, molecular strategies of biological control must be employed to gain market access. Furthermore, new biological control methods must also be convenient and widely adaptable. For commercial application, the control must be suitable for a wide range of organisms. Additionally, the ability to store the biocontrol agent for extended periods is essential. With the advent of biotechnology which can attack pests and pathogens at the most basic, genetic level, where many of these organisms have similarities, overcoming this barrier does not seem unlikely. In addition, the control method must be effective in the field over a long period of time. This entails some important characteristics. The control method must be stable in the environment and must also avoid adaptation on the part of the pathogen. Repeated use must not be necessary. Finally, any new biocontrol agent must be safe to crop plants, animals and humans; any successful alternative to chemical pesticides must be environmentally sound. It is ironic that the same molecular biological methods that have the potential to make biological control effective and useful also pose the greatest threat to the environment. The ultimate threat with the use of recombinant DNA methods is that the new genes may not be stable in the environment and may further change. Additionally, the biocontrol agent must not interfere with natural enemies, or the effectiveness of other agents, biological or chemical, in the field. Consequently, extensive testing is necessary to ensure public safety as well as effectiveness. It will be impossible to achieve zero risk, but through effective testing, the risk can at least be reduced to a level less than that for chemical pesticides.

Biological control methods have had held very little of the total pest control market and have consequently, until recently, received very little research attention. With modern advances in molecular biology, and their corresponding impact on biological control techniques, many biological control methods seem both feasible and practical. Many of these biocontrol approaches can be used in conjunction with chemical pesticides and herbicides and in many ways can remedy the shortcomings of chemicals alone. With further research and development in molecular biological techniques, and the subsequent application of this technology in biocontrol methods, it seems promising that a large number of plant pests and diseases will be managed with biological methods in the near future.

ACKNOWLEDGMENTS

Preparation of this manuscript was supported in part by National Science Foundation Grant HRD 92-53057, NASA Contract 88-357, Howard Hughes Medical Institute Grant #71194-527802 and by the W. K. Kellogg Foundation Grant # P0010125.

REFERENCES

1. Schmidt, R. R., Investigation of mechanisms: the key to successful use of biotechnology, in *New Directions in Biological Control Alternatives for Suppressing Agricultural Pests and Disease*, Baker, R. R. and Dunn, P. E., Eds., Alan R. Liss, New York, 1990, 1–23.
2. Coats, J. R., Risks from natural versus synthetic insecticides, *Annu. Rev. Entomol.*, 39, 489, 1994.
3. Weising, K., Schell, J., and Kahl, G., Foreign genes in plants: transfer, structure, expression, and applications, *Annu. Rev. Genet.*, 22, 421, 1988.
4. Russell, G. B., Sutherland, O. W. R., Hutchins, R. F. N., and Chrustmas, P. E., Vestitol: a phytoalexin with insect feeding-deterrent activity, *J. Chem. Ecol.*, 4, 571, 1978.
5. Hart, S. V., Kogan, M., and Paxton, J., Effect of soybean phytoalexin on the herbivorous insects, mexican bean beetle and soybean looper, *J. Chem. Ecol.*, 9, 657, 1983.
6. Hooykaas, P. J. and Schilperoot, R. A., The Ti-plasmid of *Agrobacterium tumefaciens*: a natural genetic engineer, *Trends Bichem. Sci.*, 10, 307, 1985.
7. Fraley, R., Rogers, R., Horsch, R., Sanders, R., and Flick, J., Expression of bacterial genes in plant cells, *Proc. Natl. Acad. Sci. U.S.A.*, 80, 4803, 1983.
8. Herrera-Estrella, L., De Block, M., Messens, E., Hernalsteens, J. P., Van Montagu, M., and Schell, J., Chimeric genes as dominant selectable markers in plant cells, *EMBO J.*, 2, 987, 1983.
9. Scnepf, H. E. and Whiteley, H. R., Cloning and expression of the *Bacillus thuringiensis* crystal protein gene in *E. coli*, *Proc. Natl. Acad. Sci. U.S.A.*, 78, 2893. 1981.
10. Schnepf, H. E. and Whiteley, H. R., Delineation of a toxin-encoding segment of a *Bacillus thuringiensis* crystal protein gene, *J. Biol. Chem.*, 260, 6273, 1985.
11. Vaeck, M., Reynaerts, A., Hofte, H., Jansens, S., and De Beukeleer, M., Transfenic plants protected from insect attack, *Nature*, 328, 33, 1987.
12. Meeusen, R. L. and Warren, G., Insect control with genetically engineered crops, *Annu. Rev. Entomol.*, 34, 373, 1989.
13. Duffey, S. S. and Felton, G. W., Enzymatic antinutritive defense of the tomato plant against insects, in *Naturally Occurring Pest Bioregulators*, Hedin, Paul A., Ed., American Chemical Society, Washington, D.C., 1991.
14. McGaughey, W. H., Insect resistance in *Bacillus thuringiensis* delta endotoxin, in *New Directions in Biological Control: Alternatives for Suppressing Agricultural Pests and Diseases*, Baker, R. R. and Dunn, P. E., Eds., Alan R. Liss, New York, 1990.
15. Gutterson N., Howie, W., and Suslow, T., Enhancing efficiencies of biocontrol agents by use of biotechnology, in *New Directions in Biological Control: Alternatives for Suppressing Agricultural Pests and Diseases*, Baker, R. R. and Dunn, P. E., Eds., Alan R. Liss, New York, 1990.
16. Oberlander, H., Advances in insect control, in *Biotechnology for Crop Protection*, Hedin, P. A., Menn, J. J. and Hollingworth, R. M., Eds., American Chemical Society, Washington, D. C., 1988.

17. Wood, H. A. and Granados, R. R., Genetically engineered baculoviruses as agents for pest control, *Annu. Rev. Microbiol.*, 45, 69, 1991.
18. Maeda, S., Expression of foreign genes in insects using baculovirus vectors, *Annu. Rev. Entomol.*, 34, 351, 1989.
19. Dougherty, E. M., Kelly, T. J., Rochford, R., Forney, J. A., and Adams J. R., Effects of infection with a granulosis virus on larval growth, development and ecodysteroid production in the cabbage looper, *Trichoplusia*, *Physiol. Entomol.*, 12, 23, 1987.
20. Hammock, B. D., Bonning, B. C., Possee, R. D., Hanzlik, T. N., and Maeda, S., Expression and effects of the juvenile hormone esterase in a baculovirus vector, *Nature*, 344, 458, 1990.
21. Kirschbaum, J. B., Potential implication of genetic engineering and other biotechnologies to insect control, *Annu. Rev. Entomol.*, 30, 51, 1985.
22. Miller, L. K., A virus vector for genetic engineering in invertebrates, in *Genetic Engineering in the Plant Sciences*, Panopoulos, N. J., Ed., Praeger, New York, 1981, 203.
23. Miller, L. K., Lingg, A. J., and Bulla, L. A., Bacterial, viral and fungal insecticides, *Science*, 219, 715, 1983.
24. Payne, C. C., Insect viruses as control agents, *Parasitology*, 84, 35, 1982.
25. Jacques, R. P., Methods and effectiveness of distribution of microbial insecticides, *Annu. N.Y. Acad. Sci.*, 217, 109, 1973.
26. Bishop, D. H. L. and Possee, R. D., Planned release of an engineered baculovirus insecticide, in *New Directions in Biological Control: Alternatives for Suppressing Agricultural Pests and Diseases*, Baker, R. R. and Dunn, P. E., Eds., Alan R. Liss, New York, 1990.
27. Davidson, E. W., Microbial control of vector insects, in *New Directions in Biological Control: Alternatives for Suppressing Agricultural Pests and Diseases*, Baker, R. R. and Dunn, P. E., Eds., Alan R. Liss, New York, 1990, 199.
28. Carlton, B. C., Development of genetically improved strains of *Bacillus thuringiensis*, in *Biotechnology for Crop Protection*, Hedin, P. A., Menn, J. J., and Hollingworth, R. M., Eds., American Chemical Society, Washington, D. C., 1988, 260.
29. Leathers, T. D., Gupta, Sc. C., and Alexander, N. J., Mycopesticides: status, challenges and potential, *J. Ind. Microbiol.*, 12, 69, 1993.
30. Hasan, S. and Ayres, P. G., Tansley review no. 23: the control of weeds through fungi: principles and prospects, *New Phytol.*, 115, 201, 1990.
31. Templeton, G. E., Weed control with pathogens: future needs and directions, in *Status of Weed Control with Plant Pathogens*, Charudattan, R. and Walker H. L., Eds., Wiley Interscience, New York, 1982, 29.
32. Burge, M. N., The scope of fungi in biological control, in *Fungi in Biological Control Systems*, Burge, M. N., Eds., Manchester University Press, New York, 1988, 1.
33. McCoy, C. W., Samson, R. A., and Boucias, D. G., Entomogenous fungi, in *Handbook of Natural Pesticides*, Vol. 5, *Microbial Insecticides*, Part A, *Entomogenous Protozoa and Fungi*, Ignoffo, C. M. and Mandava, N. B., Eds., CRC Press, Boca Raton, FL, 1988, 151.
34. Mohamed, A. K. A., Pratt, J. P., and Nelson, F. R. S., Compatibility of *Metarhizium anisopliae* var. *anisopliae* with chemical pesticides, *Mycopathologia*, 99, 99, 1987.
35. Ignoffo, C. M., Garcia, C., and Gardner, W. A., Temperature stability of wet and dry conidia of *Nomuraea rileyi* (Farlow) Samson, *Environ. Entomol.*, 14, 87, 1985.
36. Samsinakova, A. and Kalalova, S., The influence of a single spore isolate and replaced subculturing on the pathogenicity of conidia of the entomophagous fungi *Beauveria bassiana*, *J. Invert. Pathol.*, 42, 156, 1983.

37. Templeton, G. E., Weed control with pathogens: future needs and directions, in *Microbes and Microbial Products as Herbicides*, Hoagland, R. E., Ed., American Chemical Society, Washington, D. C., 1990, 320.
38. Bailey, J. A., Improvement of mycoherbicides by genetic manipulation, *Aspects Appl. Biol.*, 24, 33, 1990.
39. Kistler, H. C., Genetic manipulation of plant pathogenic fungi, in *Microbial Control of Weeds*, Tebeest, D. O., Ed., Routledge, Chapman and Hall, London, 1991, 152.
40. Gillespie, A. T. and Claydon, N., The use of entomogenous fungi for pest control and the role of toxins in pathogenesis, *Pestic. Sci.*, 27, 203, 1989.
41. Hall, R. A. and Papierok, B., Fungi as biological control agents of arthropods of agricultural and medical importance, *Parasitology*, 84, 205, 1982.
42. Kaya, H. K. and Gaugler, R., Entomopathogenic nematodes, *Annu. Rev. Entomol.*, 38, 181, 1993.
43. Georgis, R. and Hague, N. G. M., Nematodes as biological insecticides, *Pestic. Outlook*, 2, 29, 1991.
44. Georgis, R. and Poinar, G. O., Jr., Field effectiveness of entomophilic nematodes *Neoaplectana* and *Heterorhabditis*, in Integrated Pest Management for Turfgrass and Ornamentals, Environmental Protection Agency, Washington, D. C., 1989, 215.
45. Popiel, I. and Hominick, W. M., Nematodes as biological control agents: Part II, *Adv. Parasitol.*, 31, 381, 1992.
46. Poinar, G. O., Jr., *Nematodes for Biological Control of Insects*, CRC Press, Boca Raton, FL, 1979.
47. Gauglerr, R., Ecological considerations in the biological control of soil-inhabiting insect pests with entomopathogenic nematodes, *Agric. Ecosyst. Environ.*, 24, 351, 1988.
48. Georgis, R., Commercialization of steinernnematid and heterorhabditid entomopathologenic nematodes, *Brighton Crop Prot. Conf. Insectic. Fungic.*, 1, 275, 1990.
49. Stirling, G. R., Eden L., and Aitken, E., The role of molecular biology in developing biological controls for plant-parasitic nematodes, *Molecular Biology of Biological Control of Pests and Diseases of Plants*, Gunasekaran, M. and Weber, D. J., Eds., CRC Press, Boca Raton, FL, chap. 5, 1995.
50. Sayre, R. M., Biotic influences in soil environment, in *Plant-Parasitic Nematodes*, Zuckerman, B. M., Mai, F., and Rohde, R. A., Eds., Academic Press, New York, 1971, 235.
51. Olthof, T. H. A. and Estey, R. H., A nematoxin produced by the nematophagous fungus *Arthrobotrys oligospora* Fresenius, *Nature*, 197, 514, 1963.
52. Jatala, P., Franco, J., Gonzalez, A., and O'Hara, C. M., Hatching stimulation and inhibition of *Globodera pallida* eggs by enzymatic and exopathic toxic compounds of some biocontrol fungi, *J. Nematol.*, 17, 501, 1985.
53. O'Hara, C. M., Efectos de compuestos enzimaticos y exopaticos difusibles de hongos en el control biologico de nematodos, Ms. thesis, University of Peru, Lima, Peru, 1985.
54. Gonzalez, A., Jatala, P., and Franco, J., Hongos asociados con el nematodo del quiste de la papa *Globodera pallida*, Resumenes XXVII Conv. Nac. Entomol., Ica, Peru, August 1–3, 1984.
55. Mankau, R., Reduction of root-knot disease with organic amendments under semifield conditions, *Plant Dis. Rep.*, 52, 315, 1968.
56. Jatala, P., Biological control of plant-parasitic nematodes, *Annu. Rev. Phytopathol.*, 24, 453, 1986.
57. Walker, J. T., Populations of *Pratylenchus penetrans* relative to decomposing nitrogenous soil amendments, *J. Nematol.*, 3, 43, 1971.

58. Ibrahim, I. K. A., Joshi, M. M., and Hollis, J. P., The swarming virus disease of *Tylenchorhynchus martini, Int. Congr. Plant Pathol.*, Minneapolis, MN, 1973, Abstr. 555.
59. Loewenberg, J. R., Sullivan, T., and Schuster, M. L., A virus disease of *Meloidogyne incognita*, the southern root rot nematode, *Nature*, 184, 1896, 1959.
60. Christie, J. R., Biological control — predacious nematodes, in *Nematology, Fundamentals and Recent Advances with Emphasis on Plant Parasitic and Soil Forms*, Sasser, J. N. and Jenkins, W. R., Eds., Univeresity of North Carolina Press, Chapel Hill, 1960, 466.
61. Ellis, J. G., Lawrence, G. J., Peacock, W. J., and Pryor, A. J., Approaches to cloning plant genes conferring resistance to fungal pathogens, *Annu. Rev. Phytopahtol.*, 26, 245, 1988.
62. Baker, R., Mycoparasitism: ecology and physiology, *Can. J. Plant Pathol.*, 9, 370, 1988.
63. Wells, H. D., Trichoderma as a biocontrol agent in *Biocontrol of Plant Diseases*, Mukergi, K. G. and Garg, K. L., Eds., Vol. 1, CRC Press, Boca Raton, FL, 1988, 71.
64. Sundheim, L. and Tronsmo, A., Hyperparasites in biological control, in *Biocontrol of Plant Diseases*, Mukergi, K. G. and Garg, K. L., Eds., Vol. 1, CRC Press, Boca Raton, FL, 1988, 53.
65. Fulbright, D. W., Paul, C. P., and Garrod, S. W., Hypovirulence: a natural control of chestnut blight in *Biocontrol of Plant Diseases*, Mukergi, K. G. and Garg, K. L., Eds., Vol. 1, CRC Press, Boca Raton, FL, 1988, 121.
66. Sherwood, J. L., Mechanisms of cross-protection between plant virus strains, in *Plant Resistance to Viruses*, Evered, D. and Harnett, S., Eds., John Wiley & Sons, Chichester, U.K., 1987.
67. Loesch-Fries, L. S., Transgenic plants resistant to viruses, in *New Directions in Biological Control: Alternatives for Suppressing Agricultural Pests and Diseases*, Baker R. R. and Dunn, P. E., Eds., Alan R. Liss, New York, 1990.
68. Tumer, N. E., O'Connel, K. M., Nelson, R. S., Sanders, P. R., Beachy, R. N., Fraley, R. T., and Shah, D. M., Expression of alfalfa mosaic virus coat protein gene confers cross-protection in transgenic tobacco and tomato plants, *EMBJO J.*, 6, 1181, 1987.
69. Loesch-Fries, L. W., Merlo, D., Zinnen, T., Burhop, L., Hill, K., Jarvis, N., Nelson, S., and Halk, E., Expression of alfalfa mosaic virus RNA4 in transgenic plants confers virus resistance, *EMBJO J.*, 6, 1845, 1987.
70. VanDun, M. P., Bol, J. F., and Van Vloten-Doting, L., Expression of alfalfa mosaic virus and tobacco rattle virus coat protein genes in transgenic tobacco plants, *Virology*, 159, 299, 1987.
71. Templeton, G. E., TeBeest, D. O., and Smith R. J., Progress and potential of weed control with mycoherbicides, *Rev. Weed Sci.*, 2, 1, 1986.
72. Sands, D. C., Miller, R. V., and Ford E. J., Biotechnological approaches to control of weeds with pathogens, in *Microbes and Microbial Products as Herbicides*, Holland, R. E., Ed., American Chemical Society, Washington, D. C., 1990, 184.
73. Charudattan R., The mycoherbicide approach with plant pathogens, in *Microbial Control of Weeds*, Tebeest D. O., Ed., Chapman and Hall, London, U. K., 1991.
74. Tebeest, D. O., Yang, X. B., and Cisar, C. R., The status of biological control of weeds with fungal pathogens, *Annu. Rev. Phytopathol.*, 30, 637, 1992, 24.
75. Templeton, G. E., Heiny, D. K., Mycoherbicides, in *New Directions in Biological Control: Alternatives for Suppressing Agricultural Pests and Diseases*, Baker R. R. and Dunn, P. E., Eds., Alan R. Liss, New York, 1990, 279.
76. Templeton, G. E., TeBeest, D. O., and Smith, R. J., Biological weed control with mycoherbicides, *Annu. Rev. Phytopathol.*, 17, 301, 1979.

77. Yoder, O. C., The application of recombinant DNA technology to plant-attacking fungi, *N. Y. Food Life Sci. Quart.*, 15, 21, 1984.
78. TeBeest, D. O., Induction of tolerance to benomyl in *Collectotrichum gloeosporiodes* f. sp. *aeschynomene* by ethyl methanesulfonate, *Phytopathology*, 74, 864, 1984.
79. Yoder, O. C., Altered virulence in recombinant fungal pathogens, *Proc. 5th Int. Congr. Plant Pathol.*, Kyoto, Japan, 1988, 14–19.
80. Watson, A. K. and Wymore, L. A., Identifying limiting factors in the biocontrol of weeds, in *New Directions in Biological Control: Alternatives for Suppressing Agricultural Pests and Diseases*, Baker R. R. and Dunn, P. E., Eds., Alan R. Liss, New York, 1990, 305.

Index

A

Abax parallelepipedus, 76
Abiotic forces, 166
Achaea janata, 176
Agrobacterium
  radiobacter, 41, 42
  tumefaciens, 40
Agrocin 84, 41, 42
Agrocin 434, 42
Anthracnose, 161
Antibiosis, 161
Antisense RNAs. See RNAs of plant viruses
Aphids, integrated pest management, 175
Aphytis diaspidis, 178
Appressorium, fungi, insect control, 124
Arachis pintoi, 191
Arion hotensis, 76
Arthrobotrys, 73
Aschersonia aleyrodis, fungi, insect control, 127
Ascogregarina culicis, insect control, 143–144
Ascopores, 161
Avirulence genes, 161
Azospirillum, 176

B

Bacillus
  popillae, 180
  sphaericus, 74, 106
  thuringiensis, 81, 96, 106, 180, 181
Bacteria, insect control
  Bacillus
    sphaericus, 106
    thuringiensis, 106
  binary toxin, 114
  Bombyx mori, 109
  Coleoptera, 108
  CryIIIA crystal structure, 116
  Culex pipiens, 114
  delta-endotoxin, 106
  Diptera, 108
  Heliothis virescens, 109
  Lepidoptera, 108
  microorganisms, 105

mosquito larvicides, 113
  toxins, 107
  Tricoplusia ni, 109
Bacterial disease, plants
  agrocin-84, 41, 42
  agrocin-434, 42
  antibiosis, 44, 47
  avirulence genes, 43
  conjugal transfer capacity genes, 41
  copper resistance, 43
  crown gall, 40, 41, 42, 49
  Erwinia
    amylovora, 43, 44
    chrysanmi, 43, 46
    herbicola, 44
    stewartii, 43, 48
  fire blight, 44
  hypersensitive response, 43, 48
  ice nucleation, 49
  niche competition, 41, 43
  pectate lyase, 45, 46, 47
  pectinolytic enzyme, 45
  Pseudomonas
    fluorescens, 47
    putida, 47
    solanacearum, 40, 43, 47, 48
    syringae, 42
    viridiflava, 45, 47
  Saintpaulia
    ionantha, 46
    plants, 47
  soft rot, 44, 45, 46, 47
  Tra- mutants, 41
  Xanthomonas campestris, 43
Baculovirus insecticide, 91–104
  AaIT virus release, 100
  Bacillus thuringiensi toxin, 96
  co-occluded virus release, 98–99
  diuretic hormone, 93
  edcysteroid UDP-glucosyl transferase, 94–95
  environmental concerns, 97
  field testing, recombinant baculovirus, 97–100
  gene constructs, 93–96
  juvenile hormone esterase, 93–94
  maize mitochondria factor, 96

213